Contemporary Systems Thinking

Series Editor:
Robert L. Flood
Maastricht School of Management
The Netherlands

More information about this series at http://www.springer.com/series/5807

Lucia Urbani Ulivi
Editor

The Systemic Turn in Human and Natural Sciences

A Rock in The Pond

Springer

Editor
Lucia Urbani Ulivi
Department of Philosophy
Catholic University of the Sacred Heart
Milan, Italy

ISSN 1568-2846
Contemporary Systems Thinking
ISBN 978-3-030-00724-9 ISBN 978-3-030-00725-6 (eBook)
https://doi.org/10.1007/978-3-030-00725-6

Library of Congress Control Number: 2018960444

This Springer imprint is published by the registered company Springer Nature Switzerland AG
The registered company address is: Gewerbestrasse 11, 6330 Cham, Switzerland

Preface

It was as early as 1978 when Evandro Agazzi first proposed introducing into the domain of human sciences, including philosophy, the concept of "system" that cybernetics and engineering had acquired and were largely developing in their domains. Agazzi, as scientist and philosopher, could trace back to Aristotle and to the scientists operating after Galileo the early origins of the systemic idea, clearly showing that if the term "system" can be accredited with some novelty, the corresponding concept is an ancient way of looking at the things of the world that has permeated our culture since early times. But this approach was met with an almost total sunset in the early twentieth century, with general mechanicalism trailing behind metaphysical materialism, drawing the natural sciences toward a reductionist point of view and philosophy to a mostly formal and analytical attitude.

It is only fair to acknowledge that reductionism and analytical approaches have produced significant results in many fields of contemporary knowledge offering great contributions to relevant scientific discoveries and to consequent technological applications, but it is no less important to clearly limit their validity within specific fields and objectives, taking care not to improperly accredit them as the best and only methods of knowledge in every domain.

The suggested limitation has been totally overlooked, and a paradigm derived from scientific and technological success became pervasive, while the few voices observing that an individual entity escapes description in terms of its constituents, and that many problems are unsolvable if an entity is reduced to the sum of its elements, were obscured or muted without any discussion.

In time the pervasive triumphalism met with disappointment in many fields—just consider Artificial Intelligence—and it is nowadays increasingly and widely recognized that analytical or mereological approaches in ontology, and also reductionism in its many forms in epistemology, are inadequate for solving many problems and that we should introduce and support the diffusion of new concepts and different attitudes in research. The demand for "new ideas" is perceived in various fields, and finding new paths, methods, and points of view in the comprehension of our contemporary world is, if any, the proper task of philosophy.

The concept of "system," to some extent and for proper objects, is a good candidate for fulfilling such a demand, since it no longer considers the elementary constituents of an object but rather the phenomena emerging from the relations and interactions among its elementary parts.

The systemic point of view makes it possible to reconstruct several domains, both philosophical and scientific, introducing fresh ideas into research in view of a general rational vision of the world on a more comprehensive basis.

The birthdate of systemic thinking is traditionally fixed as 1967, when biologist Ludwig von Bertalanffy published his major work, *General System Theory*, opening a new attitude in research.

Nowadays systemic thinking is widely recognized and appreciated in various fields, and the production of books, conferences, and seminars testify to its vitality and conceptual richness, but its diffusion in many fields and throughout culture, in general, is far from achievement.

This book makes a return to Agazzi's program with two main focal points: firstly, contributing to the consolidation and expansion of system theoretical thinking in order to integrate the general reductionist and analytical attitude still broadly dominant in our culture and, secondly, keeping the pace with the rich and fast-growing systemic researches now expanding in different disciplines.

Since the time of von Bertalanffy's pioneering studies, systemic researchers have faced many problems, both internal and external to the systemic horizon, that were unknown or unperceived in the early days of systemics. Since then the concept of "emergence" underwent a widespread debate, and previously unexpected ideas and theoretical problems, such as those of interdisciplinarity, complexity, identity through dynamics, logical openness, quasi-systems, and incompleteness, entered onto the stage.

As a consequence many classical concepts underwent an important rethinking, including causality, abduction, objectivity, and epistemology, and also the concept of solution was revisited through dynamic and multiple approaches. All this followed in the wake of an update of traditional systemic references, to the point that we can now distinguish "first systemics," based on the concept of organization, from "second systemics," focused on coherence. The researches of Gianfranco Minati contributed greatly to the identification and distinction of the two phases of systemic thinking and warned also against the risk of "systemic reductionism."

To meet both of the aforementioned objectives, a branch of systemic research called "Systemic Researches in Philosophy, Sciences and Arts" was opened in 2009 at the Department of Philosophy of the Catholic University of Milan, hosting seminars—both public and *privatissimums*—and conferences whose results have been regularly published in dedicated sections of the *Rivista di Filosofia Neo-Scolastica* and in three volumes, all titled *Strutture di mondo. Il pensiero sistemico come specchio di una realtà complessa*, ed. Lucia Urbani Ulivi, Bologna, Il Mulino 2010, 2013, and 2015, that provide an appreciated contribution to both continuity and novelty in systemic research.

The participants—mostly well-known academicians—have been working together for many years, taking part in and holding lectures, seminars, and debates, discussing and comparing ideas through the practice of interdisciplinary work.

Presenting this book is the occasion for evaluating the many systemic activities over the last 10 years, trying to answer the questions: have the impressive contacts and exchanges of views among the representatives of different and also heterogeneous sciences merged into a new general perspective? Have the investments of intellectual energies, organization, and time been rewarded? Has all this activity been fruitful?

The investment has been rewarded, certainly as far as the public outcome is concerned, but if we ask a more subtle question, whether the many contributions and activities have been able to dismiss many prejudices and to change at least to some degree the general paradigm of our culture, then the answer is not so optimistic: the message in the bottle has been thrown in the vast ocean of shared culture, but whether somebody will collect it depends on many heterogeneous factors, mostly out of control. What John Dupré says is certainly true that culture is strongly normative in regulating our ideas and behaviors and that it is at the same time ephemeral and easily changeable, but cultural changes are hardly predictable, and ideas can stay still and steady for centuries. So we can only forward the message, and wait, and hope.

There is one last question that should be posed: which are the original contributions and new ideas likely to accredit this book as worthwhile reading?

I will briefly sketch out some relevant traits.

Vitiello and Giuliani, physicist and biologist, respectively, underline the relevance of the unobservable as the origin of the observed, linking the explicit behaviors of phenomena to hidden variables that remain opaque to direct knowledge. They suggest that there is much more to understand besides the plain data.

Matelli traces a clear conceptual link between the ancient idea of body and the contemporary concept of system, deeply rooting the contemporary in its ancient origins. She also suggests that a necessary ingredient for comprehension of ancient authors is a knowledge of the ideas, values, and objectives structuring their cultural context.

Economists Lamperti, Monasterolo, and Roventini clearly show that complexity is hardly understandable using classical models, while Fontana in architecture and Ingegnoli in bionomics underline the limits of seeking "functional optimization" in both built and natural environments.

Jurists Cafagno, D'Orsogna, and Fracchia, introducing systemic perspective into environmental legal studies, suggest viewing the environment as a common with legacy value.

Frigerio, philosopher of language, proves that a radical version of the principle of compositionality cannot be maintained, while I suggest a fresh start in philosophy of mind, considering mind and brain as different constituents of the human being, viewed as a system.

All these contributions strongly support a new manner of looking at our world: the place where the observer and the observed are defined in terms of each other, where continuous and interlaced phenomena of emergence take place, obscure to complete and precise comprehension, far from predictability, readable through principles with local validity, and open to creativity as the main resource for successfully interacting with complex and irreducible phenomena.

Whether or not such suggestions will be accepted and developed depends on many circumstances and not only on their value.

I wish to express my deep gratitude to all participants in this venture, and I especially want to mention Evandro Agazzi and Gianfranco Minati for their support and encouragement and also for freeing the philosophical debate from many inherited limitations, for opening new directions in research, and for offering solid results, which should encourage other researchers to keep the path open and constantly trodden.

Milan, Italy Lucia Urbani Ulivi

Systemic Thinking: An Introduction

Evandro Agazzi

Abstract In the 1970s, General System Theory was still the object of a radical controversy between enthusiastic supporters and fierce enemies, the firsts being attracted by the fact that GST was offering a legitimacy to concepts like those of ordered totality, global unity, goal-oriented processes, specific function, multilevel realities, and emergent properties that are frequently and profitably used in several sciences, from biology to psychology, sociology, and other "human" sciences. The enemies rejected such concepts considering them as vague, imprecise, and belonging to the superficial level of common sense language, but that should be banned from the rigorous discourse of science. This attitude was in keeping with the positivistically inspired scientific culture still predominant in the first half of the twentieth century.

The Systemic Point of View

At a distance of nearly four decades from that historical period, the systemic way of thinking has shown itself as the most proper tool for understanding complexity and investigating complex realities and stimulates reflections capable to revisit the classical philosophical concepts, the basic metaphysical and ontological principles, the deepest sense of fundamental developments in the history of science, the critical appraisal of merits, and the limitations of many present research programs in various fields.

"What are systems, what is a system?"

The answer to this question is not simple, not because this term circulates only in specialized sophisticated languages but because, on the contrary, it has acquired

E. Agazzi
Panamerican University, Mexico City, Mexico

a very large display of applications in a variety of contexts. So, for example, in ordinary language we speak of linguistic system, legal system, political system, economic system, bureaucratic system, productive system, industrial system, energetic system, railways system, metric system, and so on. In addition, within specialized disciplines one finds the mention of several systems, like muscular system, nervous system, endocrine system, and a lot of other systems in biology; numerical system, Boolean system, Euclidean system, and equation system in mathematics; or elastic system, gaseous system, and isolated system in physics, not to speak of the many systems that are considered in chemistry, crystallography, astronomy, geology, and geography and in the domain of humanities, the capital notion of philosophical system.

Such a variety of applications may at first produce the reaction of considering the notion of system as endowed with a vague and confused meaning belonging to ordinary language that can only give rise to ambiguities and misunderstandings and must be overcome and replaced by a rigorous and technical treatment. This impression is wrong, because this generalized use rather testifies that it belongs to *common sense*, that is, to that complex set of basic concepts and principles that make possible our understanding of reality and whose meaning is, therefore, not ambiguous but *analogical*, that is, such that it must be applied partly in the same way and partly in different ways to different kinds of reality. If this is the situation, the stimulating task is that of making explicit that common semantic core that underlies the analogical use and makes it possible.

This core can be described in a rather intuitive way by saying, for example, that a system is an entity constituted by parts that are linked by mutual relations, making up a complex-ordered unity which is endowed with its own individuality in the sense that it is characterized by its own properties and functioning that are different from those of its constituent parts though depending on them to a certain extent. In a shorter way, we could perhaps say that we mean by a system an ordered totality of interrelated parts whose characteristics depend both on the characteristics of the parts and on the web of their interconnections.

GST can be considered as an effort for making explicit and precise this conceptual core, by revisiting sometimes concepts and principles that philosophy has already defined and analyzed in the past but have been neglected or abandoned for several historical reasons, especially as a consequence of the predominance attained by the conceptual framework of certain successful sciences in Western culture.

The History of the Notion of System

The System of the World

The concept of system (or, better, the term "system") is so widely used in our linguistic contexts—as we have already noted—that we can be spontaneously convinced that it has belonged to our learned vocabulary from times immemorial; but if we try to retrieve in our memory of "cultivated" people, it is likely that we find its first irruption on the stage of Western culture in the title of the most famous work of Galileo, the *Dialogue Concerning the Two Chief World Systems* published in 1632, in which the reasons supporting, respectively, the Ptolemaic and the Copernican astronomic theories are compared. This occurrence, however, was not new, if we consider that already in the conclusion of his *Sidereus Nuncius* (1610) Galileo mentions as a commonly used notion that of the Copernican system and promises to present his own system later on. In short, we have sufficient evidence to maintain that the concept of system entered the background of Western culture in modernity and coming from the domain of the natural sciences.

The obvious consequence of this circumstance was that the domain of application of this concept was that studied by the new natural sciences, that is, the domain of the physical bodies considered in its generality, more or less in the sense of "external world." This is why the new science felt itself charged with the task of providing also a *system of the world* (or a "world system" for brevity), and we actually find this expression as the title of the concluding book of Newton's *Philosophiae Naturalis Principia Mathematica* (1687) that sounds *De mundi systemate* ("On the system of the world"). This phrase became quickly standard, and we find it, for example, in the title of Laplace's work *Exposition of the World System* (1796).

The System of Nature

A similar concept soon emerged in the context of another natural science, biology, and is present in the title of one of the most famous works of this discipline, Charles Linné's *Systema Naturae*, which has known, between 1735 and 1768, ten successive enlarged and revised editions during the life of its author. The significant novelty resides in the methodological structure of this work in comparison with the treatises of physics: as is well known, the most important contribution credited to Linné is the introduction of the *binomial nomenclature* for the classification of living species which, with certain improvements, has remained in use until today. For two centuries, the naturalists had been looking for a "natural criterion" for the classification of the living beings, and none had proved satisfactory. Linné's idea was that a natural classification should reflect that logical order that exists in Nature due to the fact of its being the expression of the supreme intelligence of the Creator God. In other words, the order of living creatures had to be a *logical order* for which formal logic had provided a well-known scheme in the ancient "Porfiry tree" regarding the hierarchic disposition of genera and species. According to this view, the entire Nature was conceived as a kind of mosaic in which the position of any single piece is strictly determined following a design constituted by a web of logical relations, in which each piece occupies "its" proper place. This is the fundamental worldview of *fixism*.

If we compare his "system of nature" with the "system of the world" proposed by the physicists, we can recognize that the latter was in a certain sense more significant, since the systemic architecture of the world was conceived as consisting in causal links expressed in terms of natural laws and forces, and not simply through the fragile spider web of a conceptual order, and we may also add that Linné's system has only a *descriptive* aim and purport, whereas the physicists' systems had an *explanatory* aim.

Philosophical System

All this is true, but we cannot ignore the historical background of that time, in which a prominent thinker such as Spinoza could formulate in the second book of his *Ethica More Geometrico Demonstrata* the aphorism *ordo et connexio idearum idem est ac ordo et connexio rerum* ("the order and connection of the ideas is identical with the order and connection of things"), a statement that can sound a dogmatic tenet today but that still fascinated the representatives of the romantic "philosophy of nature" at the end of the eighteenth century and also the thinkers of the German transcendental idealism of the nineteenth century. That fascination was produced by that impression of intellectual rigor, systematicity, and architectural elegance that transpires in Spinoza's work and was easily taken as a warranty of speculative soundness, as opposed to that "rhapsodic" way of thinking (to use Kant's term *rhapsodistisch*) that marks the style of those scattered reflections that ignore the need of strong logical links.

In Kant's work this kind of appraisal is explicitly made, and it is significant that the term "system" widely occurs in order to express the satisfaction of the said requirements. This explains how the phrase "philosophical system" has become customary for denoting the whole complex of the speculation of a single thinker, independently of the fact that he uses or not this denomination to this end (like does, e.g., Schelling in his *System of Transcendental Idealism*).

The Issue of Finalism

Modern natural science was born—as we have seen—according to certain ontological, epistemological, and methodological restrictions that had been proposed by Galileo and accepted by Newton and their followers. Among these restrictions one, in particular, was more implicit than explicit but of paramount importance and regarded the concept of *cause*. The "principle of causality" is one of the most fundamental metaphysical principles that can be formulated, in its simplest form, as the statement that *every change has a cause*. This principle is so fundamental that can be considered as an indispensable condition for understanding reality and, as such, universally admitted. But the concept of *cause* is far from being univocally understood, and very many meanings of this concept have been proposed in the history of philosophy. The one which is probably the most common in ordinary language is that which was called "efficient cause" in the philosophical tradition and corresponds to the idea of something that *produces* something else as its *effect*, and whose most familiar examples are the physical actions that bring about new objects or certain observable processes.

Common language, however, has no difficulty in accepting as meaningful, for example, the discourse of Socrates in Plato's *Phaedo*, when he explains that the *cause* of his coming and remaining in prison (waiting for his capital execution) was not his legs, bones, and muscles (which had served equally well to run away) but his desire to obey his city's laws. That physical situation had an immaterial cause. This depends on the fact that the Greek word *aitìa* (that is commonly translated as "cause") had a polysemous sense, which we might better express through the notion of "the reason for which." In this way we can easily understand Aristotle's doctrine of the "four causes" (formal, material, efficient, and final), of which only the efficient means the "production" of an effect, whereas the others concern the "reasons" for which something occurred, these reasons being the presence of a material substratum, the "form" or *essence* of the entity concerned, and a *goal* orienting the process. This goal can be either the *aim* or purpose pursued by an external operator performing the process or a *pattern* inscribed in the internal essence of the entity, a kind of *design* presiding over its development and also over its way of behaving in the different circumstances.

A generalized rejection of the final causes can be found at the beginning of modern philosophy, both in empiricist philosophers like Francis Bacon and in rationalist thinkers like Descartes and Spinoza. No wonder, therefore, that it enters also the new natural science, especially considering that Galileo had explicitly excluded from the objectives of this science the investigation of the *essence* of the physical bodies, and the final cause was precisely meant to reside in their essence. The same attitude is explicitly adopted by Newton who, in the *Scholium Generale* of his *Principia*, after having admitted that he had been unable to uncover the *cause* of gravitation, declares that he will not try to "imagine hypotheses," by postulating "hidden causes," like those that the Scholastic tradition was accustomed to locate in the substantial forms of things.

This kind of reasons has nothing to do with a refusal of finalism in Nature that, according to certain authors of our time, would be a subtle improper tool for admitting the interference of religion into science by requiring the existence of an intelligent omnipotent God as the cause of the marvellous order or design present in Nature. Indeed, the Newtonian statement *hypotheses non fingo* appears in the conclusion of the *Scholium Generale* in which ample space is given to a series of *theological* considerations according to which only the existence of such a supernatural spiritual Creator can account for the global order of the world, while the impossibility of uncovering the cause of gravitation is linked with the impossibility of natural science (called "experimental philosophy") to bypass the external properties of things and penetrate their intimate essence (which is an *epistemological* reason).

As a matter of fact, the mechanically interpreted order of Nature has remained for a long while one of the fundamental arguments for the existence of God as its cause, even for anti-religious thinkers like Voltaire, and the same Darwin's evolution theory (in which no finalism is present) was considered by him as the more compatible with divine creation (as he says in the final lines of the *Origin of Species* that have remained until the last edition, "There is grandeur in this view of life, with its several powers, having been originally breathed by the Creator into a few forms or into one; and that, whilst this planet has gone circling on according to the fixed law of gravity, from so simple a beginning endless forms most beautiful and most wonderful have been, and are being evolved").

System Theory

One can correctly point out that Bertalanffy's study, in which he presents the first seeds of system theory, focuses on the inadequacy of the second principle of thermodynamics for the explanation of the phenomenon of biological growth of individual organisms (*Investigations on the laws of growth*, 1934). This principle, however, had been often criticized in physics. The novelty of Bertalanffy's approach is the consideration that it applies to closed systems, whereas living organisms are *open systems*. Therefore, the problem is not that of "criticizing" the second principle of thermodynamics (that is right under its specific hypothesis) but to recognize that it is not fully *pertinent* in the case of living organisms because they are not *systems* of the kind envisaged by the said principle. Therefore, the issue is that of making a pertinent investigation regarding the different kinds of systems and possibly their common features. In particular, the question of "lawlikeness" was important, because the privilege of the physical closed systems was that of being regulated by deterministic "natural laws," permitting predictions and experimental tests, while nothing comparable appeared possible for living systems. Nevertheless, it is also evident that living organisms are able to preserve a certain identity underlying their continuous change that they realize and tend to keep a *steady state* which is different from the simple *equilibrium* (be it the mechanic or the thermodynamic one). These are among the best known characteristics that are studied in General System Theory (GST). They suggest that, due to their difference with regard to the conceptual tools usually admitted in the sciences, they can offer the opportunity of revisiting certain other more general philosophic concepts that are appreciated within the systemic way of thinking. In other words, GST, which is born in the field of *science*, can help us to recover the intellectual importance of philosophical concepts that had been marginalized as a historical consequence of the advent of modern science in the Renaissance.

Holism

The ontology of GST consists of a web of single *totalities*, each one of them being individually characterized by its own internal structure and proper functions. In order to appreciate the novelty of this ontology, it is sufficient to compare it with that of another great foundational theory in mathematics, that is, set theory. In set theory only one relation is primitive, that of membership of the elements in the set, but the elements are in a certain sense all equivalent, since they have no property. Moreover, they have no internal structure, are not even linked by particular relations, but can be arbitrarily aggregated in sets, subsets, and supersets. On the contrary, the primitive constituents in system theory are systems, each having its specific characteristics and internal structure, and they do not simply "belong" to the global system but are mutually interrelated with the other systems and are not "elements" but "subsystems" of the global system, according to a net of relations that allow the global system to have certain properties and perform certain functions.

Due to this fact, every system is at the same time "simple" (in the sense that it is well determined in what it is, independently of its relations with other systems) but also "complex" (as far as it has an internal structure, constituted by a web of relations among its subsystems, from which its own specific properties depend). It is not arbitrary nor difficult to recognize in what we have just said the classical notion of *substantial form*, which was precisely the ontological principle expressing the fact that any entity is what it is due to a particular organization of its constituent parts whose status was qualified as *matter* (not because they are simply "raw material" but because they belong to a lower level of organization). After these precisions it should be clear that "holism" is here understood as the appreciation of the "point of view of the whole" as opposed to "atomism" and has nothing to do with the notion of holism that Bunge rejects.

Complexity

In the holistic perspective, the concept of complexity is included, whose sense is that the properties of a system are the result of the correlations among the subsystems that constitute it and also of the relations it has with its environment. Modern science, on the contrary, had followed Galileo's proposals not only in the exclusion of the investigation of the intimate essence of things but also in the practice of studying an isolated phenomenon concerning one single property by trying to create an artificial situation in which all possible "disturbances" were eliminated. This is the basis of the experimental method that has given a tremendous impulse to the natural sciences and has permitted to establish numberless *physical laws* of a strictly deterministic type, from which exact predictions can be inferred. All this represents the merits of the *analytic method*.

Nevertheless, already at the end of the nineteenth century, the limitations of this approach have appeared in connection with the awareness of the impossibility of adopting this model for the treatment of *complex systems*. Nonlinearity and several forms of "indeterminism" are too well known to be recalled here. Therefore, the *synthetic approach* has emerged not at variance with, but as complementary to, the analytical approach and has produced a wide investigation on complexity that is strictly cognate with GST. This situation has promoted important philosophical discussions regarding the meaning of natural *laws* and the applicability of this concept also in other domains—like psychology, sociology, and economics—as well as a deeper analysis of the notions of determinism and causality, that is, of fundamental ontological and epistemological issues. All this is a part of more specific problems of the philosophy of science, like, for instance, the proposal of admitting as "explanation" of phenomena and also the proposal of "mechanisms" that describe "how" they occur, rather than "why" they occur, or the legitimacy of speaking of laws for single phenomena, just to mention a few examples.

Finalism

In system theory the concept of finality could receive a sense purified of any psychological flavor linking it with the intention or purpose set down by a subject (a meaning that, however, is perfectly legitimate in the study of human actions). This objective meaning of finality simply reflects the condition for qualifying something as a system, that is, the fact of being an ordered totality of parts, endowed with properties that objectively contribute, thanks to a precise order of relations and correlations (and not to another one), to the existence of properties and functions of the global system. This is actually the classical notion of "final cause," which expresses the specific way in which a certain entity behaves because it has a specific nature. If we prefer, we could say that the final cause expresses the dynamic aspect of the nature of an entity. This type of causality can be amplified also to include the supersystems of a particular system and in such a way can concern even the universe, as it was the case with Aristotle's doctrine of the "immobile motor": this acts as supreme final cause and not as an efficient cause. It is due to the Judeo-Christian doctrine of God's creation that this was also seen as efficient cause, and this—as we have already noted—produced the diffidence of certain contemporary authors against the admission of final causes in science. GST offers a conceptualization of finalism or "teleology" that is neutral and not entailing per se any "theological" consequence, though not preventing one, on the other hand, to take this finalism as an objective feature present in the world for proposing *specific philosophical arguments* for proving the existence of God, and their force must be judged according to philosophical criteria.

It may be noted, in addition, that the notion of *propensity* introduced by Popper and taken up by recent scholars for the explanation of several phenomena in the natural and especially in the human sciences is a rather patent recovery of the concept of final cause.

Interdisciplinarity

Already Bertalanffy had pointed out that the systemic approach can be applied in different domains, and this idea was strongly reinforced when the notion of an "open" system was extended not only to the existence of exchanges of matter and energy with the environment but also of *information*. In such a way, concepts like those of feedback, regulation, and self-regulation, together with all models elaborated in cybernetics, could be used for a significant improvement of the description of the interactions within systems and between systems and environment.

This means that the concept of system is *transdisciplinary*, that is, it can be profitably used in different disciplines. The systemic approach, however, is equally important in every *interdisciplinary* research that is in the treatment of *complex problems*. By complex problem we do not mean a "difficult" problem but one in which different *aspects* of an issue must be taken into consideration. In these cases the best strategy is that of making explicit the differences and specificity of the disciplines that can approach each aspect, with their specific criteria of investigation, of testing, and of making arguments, and then to make the effort of making a certain translation and especially of finding correlations between these disciplinary results. The "global" result will not be, and must not be, a "unique" portrayal of the reality investigated (obtained by *reduction* to a single allegedly "fundamental" discipline) but a multifaceted portrayal in which the contribution of every discipline can be appreciated because it "contributes" to a better understanding *of the whole*.

Considering the enormous quantity of complex problems that are surfacing in our contemporary world, and which will increase in number and complexity in the coming future, we can conclude that a generalized adoption of a systemic way of thinking will be the more suitable intellectual attitude to be promoted in our societies.

Mexico City, Mexico Evandro Agazzi

Contents

Contributors

Evandro Agazzi Panamerican University, Mexico City, Mexico

Maurizio Cafagno Dipartimento di diritto, economia e culture, Università degli studi dell'Insubria, Como e Varese, Italy

Domenico D'Orsogna Università di Sassari, Sassari, Italy

Carlotta Fontana Dipartimento di Architettura e Studi Urbani—DAStU, Politecnico di Milano, Milano, Italy

Fabrizio Fracchia Università Bocconi, Milano, Italy

Aldo Frigerio Department of Philosophy, Università Cattolica S. Cuore, Milan, Italy

Alessandro Giuliani Istituto Superiore di Sanità, Rome, Italy

Vittorio Ingegnoli Department of Environmental Sciences and Policy—ESP, Università Statale di Milano, Milan, Italy

Francesco Lamperti Institute of Economics, Scuola Superiore Sant'Anna, Pisa, Italy

Elisabetta Matelli Dipartimento di Filologia classica, Papirologia e Linguistica storica, Università Cattolica del Sacro Cuore, Milan, Italy

Gianfranco Minati Italian Systems Society, Milan, Italy

Irene Monasterolo University of Economics and Business (WU), Vienna, Austria

Andrea Roventini Institute of Economics, Scuola Superiore Sant'Anna, Pisa, Italy

Lucia Urbani Ulivi Department of Philosophy, Università Cattolica del Sacro Cuore, Milan, Italy

Giuseppe Vitiello Dipartimento di Fisica "E.R.Caianiello", Università di Salerno and Istituto Nazionale di Fisica Nucleare, Salerno, Italy

Phenomenological Structural Dynamics of Emergence: An Overview of How Emergence Emerges

Gianfranco Minati

Abstract We propose a conceptual overview of the phenomenology of emergence, dealing with some of its crucial properties, representations, and specific inducing phenomena. We focus on properties such as compatibility and equivalence, and their interplay, as a basis suitable for hosting and inducing processes of emergence. We specify this interplay by considering suitable hosting processes, such as synchronisation, covariance and correlation, coherence, and polarisation. We then consider phenomena where such processes are considered to occur, providing suitable foundations for the establishment of processes of emergence, such as the establishment of attractors, bifurcation points, chaos, dissipation, domains of coherence, multiple and remote synchronisations, and multiple systems. We list properties of representations understandable as signs, clues, and possible trademarks of the coherence of interplaying compatibilities and equivalences. This interplay establishes processes of emergence, such as the presence of bifurcations, meta-structural properties, network properties, non-equivalence, power laws, scale invariance, symmetry breaking, unpredictability, and the constructivist role of the observer. Such interplay is considered in the continuing absence of a consolidated theory of emergence and within a new, generalised conceptual framework where theorisation is no longer considered as a necessary perspective. Finally, we briefly discuss issues relating to simulation.

1 Introduction

The purpose of this chapter is to present a general, inevitably partial overview of the phenomenology of emergence and considers some related properties and representations, the interplay between compatibilities and equivalences, and phenomena as suitable incubators of types of emergence (see Sect. 2).

G. Minati (✉)
Italian Systems Society, Milan, Italy
e-mail: gianfranco.minati@AIRS.it

L. Urbani Ulivi (ed.), *The Systemic Turn in Human and Natural Sciences*,
Contemporary Systems Thinking, https://doi.org/10.1007/978-3-030-00725-6_1

We consider the interplay among suitable hosting processes, such as synchronisation, covariance and correlation, coherence, and polarisation, the establishment of attractors, bifurcation points, chaos, dissipation, domains of coherence, multiple and remote synchronisations, and multiple systems, and properties such as bifurcations, meta-structural properties, network properties, non-equivalence, power laws, scale invariance, symmetry breaking, unpredictability, and the constructivist role of the observer (Sects. 4 and 5).

The overview that we present is mainly conceptual and intended for *interdisciplinary usage*. This includes recognising predominant properties, processes and their combinations, as well as suitable approaches, correspondences, representations, and transpositions when problems and solutions of a discipline give rise to or constitute problems and solutions for another discipline. **This contributes, with reference to the topic of the book, to the *transdisciplinary* study of systemic properties when systemic properties are studied *per se* and not systems having such properties, e.g. in biology, economics, linguistics, and physics**.

This overview is presented in the persisting absence of a consolidated theory of emergence and in a new generalised conceptual framework where theorisation is no longer considered as a necessary perspective, "contenting" ourselves with concordances and correspondences in a data deluge (Anderson 2008) *such as the properties listed in Sect. 5.2*. However, the approach assuming that *correlation supersedes causation and theorising is mathematically wrong* (Calude and Longo 2016).

More realistically, we should consider *soft-theorisations, quasi-theories* as partial, regarding composite phenomena of the global system under study. For instance, in the study of emergent collective behaviour, e.g. flocking and traffic, we have available theories of fluid dynamics, gravitation, thermodynamics, and topology, however insufficient, to constitute a theory of collective behaviour. In the same way, we may consider the intrinsic limitations of the theory of phase transitions to model phenomena of emergence (Minati and Pessa 2006, pp. 229–230). Section 5 considers how *incompleteness* for theorisations may be considered related to the impossibility to fully *zip* the representations of system behaviour into analytical formulae.

Theories are considered here as corpuses constituting explicit symbolic approaches contrasted with non-explicit, non-symbolic approaches such as networks and sub-symbolic approaches. A theory is considered as a coherent group of tested or verifiable general propositions allowing extrapolations and *falsification*.

Examples are the Big Bang theory, currently the prevailing cosmological model for the birth of the universe; Darwin's theory of evolution; Quantum Field Theory; and the Theory of Relativity.

Examples of alternatives are data-driven approaches, i.e. to cluster retrospectively by finding emergent correspondences without looking for the respect of theoretically pre-established ones, by using, for instance, statistics, finding correlations, or by using neural-network-based models.

The concepts introduced should be useful for dealing with problems of complexity, particularly where such problems are not well formalised and may never be so, for instance, in architecture, economics, education, medicine and welfare, philosophy, political sciences, safety at work, and other general social issues. We mention how social systems, problems of post-industrial society, where knowledge is the main resource and which is characterised by significant levels of complexity, are often still dealt with by using only knowledge from pre-complexity industrial society (Kumar 2004; Minati 2012a, b), i.e. ignoring processes and properties of complex phenomena.

While several robust, although partial, formalisations are available for the study of complexity (see Sects. 4 and 5), the interplay of equivalences, recovery dynamics, and other properties conceptually introduced here may need to wait to be properly formalised within the perspective of further research.

Various interdisciplinary contributions have introduced and elaborated the concepts of:

- Self-organisation, considered here as a continuously variable, but stable process of acquisition of new structures, e.g. either periodic, quasi-periodic, or quasi-predictable. Examples include swarms having repetitive behaviour and dissipative structures such as whirlpools, the Bènard rolls (Getling 1998), and structures formed in the Belousov-Zhabotinsky reaction (Kinoshita 2013). Stability of variability, e.g. periodicity, corresponds to stability of the acquired property.
- Emergence, considered here as continuous and irregular, but coherent, e.g. dynamically correlated, variable in the acquisition of new structures and non-equivalent processes of self-organisation. Examples include the continuous, irregular, and unpredictable, but coherent acquisition of shapes by swarms and traffic distributions. Some contributions presented suitable models and, in some cases, simulations (see, for instance, Anderson and Stein 1985; Batterman 2011; De Wolf and Holvoet 2005; De Wolf et al. 2005; De Wolf et al. 2006; Fernandez et al. 2014; Krause and Ruxton 2002; Licata and Minati 2016; Manrubia and Mikhailov 2004; Samaey et al. 2008; Vicsek and Zafeiris 2012).

Returning to the theme of this contribution, we specify that self-organisation and emergence are phenomena intended as being continuously established by related constituting processes, e.g. sequences of dynamical coherent aggregations. The concept of coherence may be considered coincident, for the moment, with correlation. The concept of coherence is elaborated in Sect. 4.3.

Different levels of representation of phenomena of self-organisation and emergence can be considered as microscopic, macroscopic, and mesoscopic. Microscopic considers each specific agent as a well-distinguishable entity, such as boids, molecules, or customers in markets. Macroscopic considers global variables ignoring any specificities of indistinguishable agents, such as patterns, temperature, or daily revenue in markets. Mesoscopic considers properties of clusters of microscopic agents without ignoring them completely. This could include instantaneous clusters of boids, possibly spatially dispersed but flying at the same altitude; molecules clustered by same energy level and chemical properties; numbers of cars in traffic that cannot accelerate (we cluster cars slowing down, cars at constant speed in the queue, and cars standing still in a queue), or daily revenue subdivided by product and time period of sale in markets. This is consistent with the definitions of the three levels described by Liljenstrom and Svedin (2005). We consider how processes of emergence may occur in different equivalent ways.

We consider here collective phenomena established by processes of emergence intended as *mesoscopic coherence* (Minati and Licata 2013). Clusters considered by the mesoscopic representation are based on microscopic equivalences among clustered elements. We elaborate here the interplay between such equivalences as a "mechanism" leading to coherences and processes of emergence.

The structure of this contribution is as follows:

- The first part summarises the basic properties of complex systems and emergence, introducing the central role of the coherent interplay of equivalences and compatibilities.
- The second part discusses crucial aspects of complex systems and their generative processes of emergence.
- The third part briefly describes some generic processes and their properties suitable for hosting the interplay between compatibilities and equivalences. This includes synchronisation, covariance and correlation, polarisation as global ordering, and relationships between correlation and coherence.
- The fourth part considers phenomena where processes mentioned in the third part are considered to occur and provide suitable places for the establishment of processes of emergence. This includes *phenomena* such as the establishment of attractors, bifurcation points, chaotic behaviour, dissipation, domains of coherence, multiple and remote synchronisations, and multiple systems. Furthermore, we list *properties of representations* of processes considered as signs, clues, and possible trademarks of coherence of interplaying compatibilities and equivalences establishing processes of emergence, such as having attractors, bifurcation points, meta-structural properties, network properties, non-equivalence, power laws, scale invariance, symmetry breaking, unpredictability, and finally the constructivist role of the observer.
- Section 6 briefly discusses issues relating to simulation.

2 Emergence from Compatibilities and Equivalences

Among the various possible processes of emergence (see below) consider a sufficiently representative prototype: the case of the establishment of collective behaviours or collective motions in *3D* space by generic agents. For example, living agents assumed to possess a suitable cognitive system (boids and fishes); living agents assumed to possess no cognitive system (amoeba, bacterial colonies, cells, and macromolecules); non-living agents (lasers, networks of oscillators, traffic signals, or rods on vibrating surfaces). For a more detailed overview see, for instance, Vicsek and Zafeiris (2012). When dealing with collective behaviour, in the following we also use both the terms "component" and "entity" to denote a composing, belonging generic agent.

The following considerations may be suitably generalised across disciplines thanks to the genericity of the agent considered and by considering other *nD*, non-Euclidian, and possibly non-physical spaces having properties such as spaces of values, e.g. prices, or of possible events.

As stated above, we consider here the case of spatial emergent collective behaviours given by temporally subsequent coherent changes occurring in *3D* space such as for flocks, swarms, traffic, and possible related simulations (Vicsek and Zafeiris 2012). We stress the phenomenological importance of the *homogeneity* of agents giving rise to the collective behaviour under study such as boids forming flocks, mosquitoes forming swarms, fishes forming schools, firms forming industrial districts, oscillators forming networks of oscillators (Lind et al. 2009), and signals or vehicles forming traffic phenomena. The homogeneity of the constituent components ensures *sameness*, for instance, of rules of interaction, parameters, ranges of values for variables, and of the ways to process (cognitively or not) environmental inputs. We do not consider here any possible cases of non-homogeneous, i.e. mixed, or unlikely collective behaviours.

The peculiarities of emergence include, on the one hand, robust coherence given, for instance, by the occurrence of long-range correlations (see Sects. 4 and 5) among components and, on the other, intrinsic low predictability, singularities, with little or no possibility of external regulation. Such peculiarities are intended here as being related to the dynamic interplay among compatibility and equivalence, as specified below, of multiple, instantaneous, subsequent microscopic roles, e.g. spatial positions, velocity, and direction, of components or of their configurations forming a collective behaviour. This is suitable, for example, for the case of the mesoscopic level which is considered as the place of continuous *negotiations* between the micro and macro. At the mesoscopic level, a large variety of equivalent mesoscopic representations are possible because of the undefined number of possible multiple clusters (Laughlin et al. 2000), such as, for example, Multiple Systems and Collective Beings (Minati and Pessa 2006) and Networks (see Sect. 5).

Interactions among agents may occur in several ways. A *first simplification*, the simplest case occurs when considering couples having a single and completely identifiable interaction between each component of the couple. In this first simplification, the interaction in progress is not considered as being composed together with other superimposed interactions having, for instance, different durations and intensities. ***However, depending on the granularity of time considered, at each instant we may have several interactions in progress regarding different couples, independent of their initial times and durations***.

The approach which considers the states of components per instant of discretised time as *resulting and not in progress* is a *second simplification*. Interfering processes which are differently temporally autonomous, i.e. have different initial times, durations, and intensities, should be considered as progressively establishing the collective behaviour. In models and simulations (see Sect. 6), this simplification usually comes together with the first simplification when considering single non-interfering temporally subsequent interactions. However, real cases of populations of *interfering* interactions (for instance, occurring in physics as a superimposition of waves and disturbances, where the interference changes the effects of interactions or the interaction itself) having different initial times, durations, and intensities, differently superimposed are analytically intractable. The problem calls for statistical and macroscopic approaches which however hides and does not consider the properties of interactions, as discussed in Sects. 4 and 5.

From here we assume the validity of the two conceptual simplifications mentioned above relating to single interactions and time discretisation. We use such simplifications to introduce conceptual commentary on compatibility and equivalence. The mesoscopic level will, however, allow the consideration of clusters of analytically intractable aspects (see, for instance, Sects. 4.3 and 5.2) without completely losing the microscopic aspects.

Within the conceptual framework outlined by the previous considerations, we then take into account how the dynamics of agents occur as selections among compatible and equivalent possibilities.

2.1 Admissible, Compatible, and Equivalent

Time can be considered discretised in $h\,\Delta_t$ instants (the granularity of time), such as for image refresh rates used for monitors, regular identical time sequences in simulations, or when considering data resulting from processes of scanning images at different levels of quality (differentiating between photo- and text-quality scanning). In this context sequences of temporally subsequent changes per instant of interacting agents or generic components of a collective behaviour, should be both admissible and compatible as introduced below. Considering values taken by a specific variable, e.g. speed, microscopically related to each generic agent at time t_h and its subsequent value at time t_{h+1}, the latter must be:

- **Admissible**, that is respecting general physical evolutionary constraints of the phenomenon under study. This is required for any subsequence. For instance, the subsequent spatial position of the same agent cannot differ from the previous one more than that given by a maximum admitted speed, and agents cannot appear and disappear (we assume here the validity of classical physics).

Admissibility is a microscopic property of single agents related to changes considered possible regardless of changes in other agents.

Admissibility is of a microscopic nature.

We consider now **compatible** changes.

Compatibility is intended as a *local* property between the values of the neighbourhood, e.g. metrical, topological, or energy levels, of each agent at a subsequent time t_{h+1}. Such a neighbourhood is considered here as being constituted by other agents, neglecting the environment. More generally, in the framework introduced above, admissible microscopic instantaneous possible changes may be:

- Non-**compatible** because positional changes (e.g. spatial) may lead to crashing trajectories or occur in such an inhomogeneous way as to be inconsistent with the coherence of the collective behaviour. For example, allowing unacceptable changes in densities and braking anisotropy (the property of being directionally dependent). A specific microscopic state may have several different possible and compatible changes available, all allowed to occur. However, admissible and compatible sequences are not necessarily coherent, i.e. they may not necessarily establish a collective behaviour, for example, if their changes are poorly or contradictorily correlated. Furthermore, compatible changes may constitute, for instance, numerable sets or local spaces (e.g. metrical) and may have different properties such as in their topology.

The closeness of neighbourhood (see Sect. 4) is usually intended as having metric relationships (Reynolds 1987) between the agents under consideration, i.e. the space radius considered as fixed or context-dependent in simulations. However, the closeness may be based on the topological distance considering how many intermediate agents separate two specific agents when the intensity of interaction does not decay with an increase in the metric distance. In this case, intermediate agents convey *mediated* information allowing medium range interactions (Ballerini et al. 2007). In summary, one spatial position may be, for instance, metrically or topologically compatible with others.

Compatibility is considered at a neighbourhood, e.g. metric or topological, level.

Such an idea of compatibility could, in the abstract, allow the understanding of collective, coherent behaviours as improbable *selections* among compatible instantaneous predetermined configurations of neighbourhood agents.

Another improbable possibility is to consider agents computing, i.e. deciding, the next change *sequentially*.

In this view at each instant, only a single agent would be authorised to change, in a step-by-step digitalised mode, unless one allows parallel computing, leaving however unanswered how sequences and turns are decided.

Similar to mathematics where a set of simultaneous equations is solved together, a more realistic possibility consists of considering dynamical instantaneous neighbourhoods continuously computing, i.e. deciding, local compatibilities. This would allow a possible continuous process of *tolerant* balancing among multiple neighbourhoods established by the same belonging agents (see below for a discussion about tolerance percentages). It is a matter of dynamical collective maintaining or, better, recovering of neighbourhood compatibility at different levels avoiding disintegration. *We consider instantaneous global coherence as being given by instantaneous local coherences which, realistically, occur at different levels and percentages as considered below.*

Within the framework introduced above, we consider now the possible **equivalence** of compatible microscopic instantaneous possible changes at a subsequent time t_{h+1}.

Consider, for instance, the case of a swarm of mosquitoes flying around a light. At each instant, each mosquito has available a selection of possible compatible and equivalent positions, some having the same collective effects on the coherence of the swarm.

The same applies to subsequent instantaneous compatible and equivalent sequences of emergent *configurations* of the swarm.

Similar is the case of sequences of frames which constitute a movie. Every moment there are subsequent possible equivalent frames such that replacing one with another does not disturb the overall emerging sequence, i.e. the perception and the meaning of the movie does not change for the observer.

At first, the equivalence of possible compatible changes in the values of single variables, e.g. spatial, economical, and chemical, may be considered due to their *irrelevance* for the process of emergence. For example, we may consider possible changes as locally equivalent, when closeness may be metrical or topological as introduced above. Moreover, irrelevance may relate to the irrelevance of effects. Such irrelevance must be better specified, for instance, belonging to ranges of values, being solutions of equivalence functions or, possibly, by using fuzzy logic.

Equivalence of changes may be given by the *inter-changeability* allowed by irrelevance, when their differences do not affect emergent properties such as patterns or behaviours. In this case, we consider the equivalence of changes contributing to general coherence, e.g. contributing to the scale invariance and paths of attractors (Sect. 5). Local equivalence of neighbourhoods occurs when there is a stability of properties such as ordered sequences, topological roles, e.g. being on the edge or at the centre, and density.

Equivalence may then relate to maintaining the same irrelevance towards the global properties of the collective behaviour.

More generally, we may consider equivalence as occurring when possible changes are *irrelevant* to the consistency of the coherence of the collective behaviour. *In short, there is equivalence when one change is worth the other*. For instance, as we will see in Sect. 4, there are different equivalent ways for single agents to maintain scale invariance and power laws, responsible for maintaining coherence in collective behaviours.

At this point, we may consider that the more equivalent choices which maintain properties are available, the higher the probability that the collective behaviour is maintained (numbers of equivalences available may be considered in correspondence to levels of robustness). For living systems, maintaining equivalence may also be intended as a strategy against predators whose decision-making is confused by large equivalences, and as a strategy to minimise the individual possibility of being predated (see, for instance, Duncan and Vigne 1979; Foster and Treherne 1981; Magurran 1990; Pulliam 1973; https://www.revolvy.com/main/index.php?s=Antipredator%20adaptation&item_type=topic). This is explored in the more general conceptual framework of Sociobiology as introduced by Edward Wilson (1975).

The availability of large quantities of equivalences facilitates, in general, emergence allowing coherence rather than uniqueness.

We stress that we are considering possible changes as equivalent choices in a given instant. However, equivalent changes at time t_h may be *evolutionary* non-equivalent, giving rise to subsequent non-equivalent evolutionary pathways. This is the case for chaotic behaviours characterised by high sensitivity to different initial conditions which may seem, at the initial stage, equivalent choices to the system.

Starting from a real configuration of agents at time t_h, we may hypothetically consider among all possible subsequent t_{h+1} compatible changes, the subsets of those maintaining coherence. This allows us to consider the set of all possible equivalent configurations of changes.

For example, we may consider compatible spatial positions as phenomenologically equivalent as long as they do not generate strange topological configurations, unreasonable changes in density, or do not occur through abnormal fluctuations.

Furthermore, the coherence considered above may be detected in different ways depending on the interest of the observer (see Sect. 5) and on the representation considered.

As an example, we may consider compatible spatial positions as analytically equivalent until they do not change or destroy correlation (as described in Sects. 4 and 5) and therefore the general coherence. *In other words, equivalence relates to the fact that there are equivalent ways to generate or maintain the same coherence.*

In the abstract, we may consider the sets of equivalent changes available per instant to single agents and the network of mutual constraints of interdependence among them maintaining general coherence.

Selection by an agent of a state among all equivalent ones available to it reduces the number of equivalent possible states for other agents.

In reality, the dynamics of coherent complex behaviours assumes the *continuous computation* of equivalences as in the case of simultaneous equations considered above.

Properties of spaces (such as topologies, and of sets, such as quantities of their elements, their changing with possible regularities and interdependences) constituted by equivalent changes available per instant should be explored by further suitable research looking for possible correspondence with evolutionary properties of the collective behaviour. Such properties will allow one to consider *types of incompleteness and their classification*.

However, in reality, we face instantaneous variable combinations of adopted changes at time t_{h+1} having different percentages of equivalent and non-equivalent changes. We also face variable temporal sequences of such instantaneous combinations, e.g. per level of percentages.

Properties of the distribution of such sequences of percentages will correspond to and represent the robustness of the collective behaviour under study. A behaviour is *robust because it is tolerant to percentage changes*, with robustness being related to the ability to recover temporary *tolerable* (i.e. non-catastrophic) non-equivalences. *We should build an understanding of continuous processes of local and general emergence, their possible levels of dissolution, propagation, and composition taking place through equivalences* (Longo and Montévil 2014; Paperin et al. 2011) *and multiple roles of the same components.*

Further research should focus on the quantitative dynamics of temporary sets of equivalent and non-equivalent changes. The problem may be considered as the study of oscillations of the instantaneous number of agents respecting a critical property, such as scale invariance, power laws, and paths of an attractor, around the total number of agents involved.

While episodic critical percentages of non-equivalence may be overcome, their significant re-occurrence in sequences will possibly initiate irreversible disaggregation of the collective behaviours by propagating a decrease in the number of equivalences.

We should expect variable domains of possible changes in coherence. Properties of such domains include their continuity or topology, and their meta-stability—the ability to maintain or switch between states due to small fluctuations (see Christensen and Moloney 2005). Correspondences with properties of processes of emergence, such as linearity or establishing singularities, should be a target of further research.

Different levels and combinations of equivalences should be intended as structural degrees of freedom for the collective behaviour under study.

2.2 Several Ways to Maintain Emergent Coherence

Properties of the distribution of sequences of percentages of equivalent and non-equivalent changes are intended to represent the dynamics of the coherence. However, other properties may represent instability, degeneration, or lack of coherence tout-court. Furthermore, they may also represent the actions of configurations as initiators, founders of different non-equivalent paths (i.e. different histories as different coherences). Some configurations may give rise to a new set of coherent dynamics (non-equivalent to the previous one), e.g. the changing of collective behaviour from circular to directional. Alternatively, they may give rise to a subsequent non-coherent behaviour leading to dissolution of the previous coherence. Furthermore, combinations are possible when, after local or general dissolution, collectively interacting agents resume local or general coherence such as for a flock temporarily disaggregated by external perturbations of any nature.

The dynamical balancing between equivalences and non-equivalences relates to meta-stabilities and dynamical order as in structural dynamics (see Sect. 4.3).

Hypothetical global coherence should be interpreted as a sequence of dynamical local coherences, correlated by long-range, scale-free correlations (see Sect. 4).

Here, we consider properties of compatibility and equivalence as suitable to properly represent the crucial dynamics of processes of emergence.

The crucial question: "how compatible and equivalent *options are in reality selected by agents*", in the case of non-living agents, may be approachable by considering the suitability of rules of interactions, parameters, processes of meta-stability, energy reasons, and their networks (see Sect. 4).

In the case of living agents possessing a cognitive system supporting decision-making, the question is more difficult. In Sect. 2.1, we mentioned how *selection is not sequential or optimised but, rather, constantly collective*. Simplified, reductionist approaches, such as considering processes of optimisation, are as inadequate as the optimal choice criterion in economics to model the process of choice in markets (Cartwright 2014). We consider the comment "How starlings achieve such a strong correlation remains a mystery to us" (Cavagna et al. 2010, p. 11869) to be possibly generalisable to generic collective behaviours, although processes of synchronisation, as introduced in the latter, may be explanatory for some specific cases. *We focus here* (see Sects. 4 and 5) *on representations of properties and the effects of selections* on equivalence.

Equivalence is considered here as a *source* (reason) of microscopic unpredictability, incompleteness, and at the same time of stability, or robustness of the collective behaviour being allowed to occur in different though equivalent ways.

2.3 Incompleteness, Fluctuations, and the Observer

The interplay among equivalences takes place within the conceptual framework of *theoretical incompleteness* (Minati 2016a; Minati and Pessa 2018, Chap. 4) defined, in part, by:

- *Non-decidability* when there are conceptually no *algorithms* to automatically decide, i.e. there is no algorithm which produces the corresponding solution in finite time (Turing 1936);
- The *principles of uncertainty* for which, in physics, the search for increasing accuracy in measuring the value of one variable involves a reduction in knowing the value taken by another. This is the well-known Uncertainty Principle (Heisenberg 1971).
- The *complementarity principle* for which some physical objects have complementary properties which cannot be observed or measured simultaneously. This is the case of corpuscular and wave aspects of a physical phenomenon (Bohr 1928).
- *Incomplete, non-explicit. and non-univocal modelling* as in the notion of DYnamic uSAge of Models (DYSAM) related to situations in which the system to be studied is so complex that it is impossible, in principle, to fully describe it using a single model nor a fixed sequence of models (Minati and Pessa 2006, pp. 64–75).
- *Non-computable uncertainty*, i.e. non-computable probability given, for instance, by uniqueness and singularities (Bailly and Longo 2011).
- Impossibility to *fully zip* the representations of system behaviour into analytical formulae, i.e. use *ideal models*, see Sect. 5.
- *Sloppiness*. "Sloppy is the term used to describe a class of complex models exhibiting large parameter uncertainty when fit to data" (Transtrum et al. 2015, p. 2). Theoretical sloppiness refers to models which exhibit behaviour controlled by a relatively small number of parameter combinations such as in physics, biology, and other disciplines. The theme of sloppiness refers to the fact that, within certain limits, the precision is deleterious (Anderson 2008).

- Quasiness. The concept of quasi is used in diverse disciplines, such as for quasi-crystals, quasi-particles, quasi-electric fields, quasi-periodicity, and quasi-systems, concerning modelling approaches such as network models, in this case called quasi-networks (see below). In general, the concept of quasiness for systems concerns their continuous structural changes which are always meta-stable (Minati and Pessa 2018).

Such dynamical interplay among equivalences and its properties, such as incompleteness, provide suitable *places* for the establishment of processes of emergence. *Incompleteness is probably one, if not a necessary though non-sufficient condition for the establishment of processes of emergence.* With regard to this, we present in the following some general conceptual frameworks to be updated by possible future research. Future research should consider the introduction of possible supplementary, necessary and sufficient conditions for the occurrence of processes of emergence.

However, such dynamical interplay should achieve an unavoidable conclusion due to, for instance, decisive fluctuations (deviations of the actual time evolution of an observable from its average evolution, Uzunov 2010) within a system subject to random forces or undergoing chaotic motion (see Sect. 5.1).

Nevertheless, emergent systems are consistent, robust, they maintain their identity and continuously acquire new coherent consistent properties rather than the *same* properties (as for non-complex systems, see Sect. 3). A possible related question is "when does a collective behaviour cease to be such, e.g. when is a swarm no longer a swarm"?

This chapter intends to contribute to the conceptual clarification and, possibly, better specification of *ways* by which generic processes of emergence take place, and to the better understanding of the possible properties of such ways (Cruchtfield 1994). **Everything, including processes of emergence, *must happen in some way, or even in multiple and equivalent ways*. Furthermore, the ways considered here constitute significant (we cannot say necessary or sufficient) signs, clues, ingredients, *theory-substitutive* concordances, and correspondences, and possible trademarks of coherence of interplaying compatibilities and equivalences establishing emergent processes and properties. S*uch processes cannot be considered as having been reduced to single ways through which they occur or by ignoring the dynamics of coherence*, as intended in a reductionist view (see Sects. 4 and 5**). General equivalences, incompleteness, and multiplicity are contextual properties making Lego-type approaches unsuitable in principle.

Finally, we must underline the role of the observer (Fields 2016). This role is not reduced to a cause of relativism (See section "The Constructivist Role of the Observer"), but, rather, to a source of cognitive existence such as for constructivism and ontologies (Baas 1994; Heard 2006) and abduction (Minati and Pessa 2006, p. 54). It is a matter of cognitive representation and the detection of properties, such as when re-emerging at different levels of scaling and representation, for instance, in the case of dissipative structures and their emergent properties occurring on the scale of whirlpools and hurricanes.

3 Complex Systems and Emergence

Briefly, complex systems differ from non-complex systems in that they are continuously and coherently acquiring new properties over time rather than possessing the same *iterated* properties. *Sameness is replaced by coherence.*

Simple examples of usually *non-complex systems* include electrical networks (power grids), automata (mechanical and electronic systems such as cars and televisions), hydraulic systems (aqueducts or domestic heating systems), assembly lines, and procedural systems (sequences in the assembly of an engine or online booking procedures).

Simple examples of *complex systems*, systems where there is occurrence of processes of emergence, include autonomous lighting networks which tend to adopt coherent variations, such as communities of coherently light-emitting bugs, e.g. fireflies, and autonomous lighting networks (Minati et al. 2016); road and rail traffic acquiring properties such as queues or delays; cellular automata; chaotic systems (e.g. the climate); dissipative structures (whirlpools in fluid dynamics); double pendulum; flocks; industrial district networks; traffic signals (Internet acquiring properties such as vulnerability); swarms; social systems (cities, schools, hospitals, companies, families, and temporary communities, such as passengers and audiences), and telephone networks.

A distinction (Longo and Montévil 2014) can be made between self-organisation and emergence (De Wolf and Holvoet 2005; Pessa 1998) (see the Introduction above), the first being when a sequence of new properties is acquired in a phase-transition-like manner (Minati and Licata 2012), and has regularities and repetitiveness, such as a swarm around a light, or synchronisations.

Emergence (Batterman 2011; Heard 2006; Pessa 2006) **is considered as the continuous** (Paperin et al. 2011)**, irregular, and unpredictable process of acquisition of non-equivalent, i.e. non-deducible from one another, compatible,** ***coherent*** **sequences of new properties establishing a complex system.**

The generic process of emergence is considered here as the selection, through fluctuations, of one configuration among all possible subsequent compatible new configurations which can all occur with coherence being maintained. This selection through fluctuations among **equivalent,** ***since*** **compatible and coherent,** ***possibilities, is at the root of the radical unpredictability of emergence.***

The sustainability of processes of emergence should also be considered. There are a large variety of possible reasons why a process of emergence may be extinguished, such as a decrease in density weakening spatial interactions, a lack of available energy leading to the weakening of interactions between entities, or a loss of coherence, for example, because of dominating external noise.

4 Synchronisation, Correlation, Coherence, and Polarisation

In this Section, we briefly list some generic processes suitable for hosting the interplay between compatibilities and equivalences: conceptual places suitable for the occurrence of processes of emergence establishing complex phenomena.

4.1 Synchronisation

The classical concept of ***synchronisation*** in physics relates to oscillatory phenomena, such as individual oscillators, being *in phase*. The concept of synchronisation has various disciplinary meanings, such as referring to maintaining the *same* parametrical value in oscillatory phenomena, e.g. alternating current power, marching parades, and swinging pendulum.

In this chapter, synchronisation refers to the establishment of stable phase relationships among a population of oscillating components (Acebrón et al. 2005; Hong et al. 2005) such as communities of fireflies (Buck and Buck 1966).

The concept of synchronisation applies to complex phenomena such as synchronisation of chaotic systems (Boccaletti 2008; Boccaletti et al. 2002), networks of chaotic oscillators (Minati et al. 2015), and autonomous lighting networks (Minati et al. 2016).

4.2 Correlation

In statistics and probability, the concept of correlation is very similar to that of *covariance* (Doncaster and Davey 2007), both measuring a certain kind of dependency between the variables under study. More precisely, while variance (Roberts and Russo 1999) measures how far a set of random numbers deviates from their mean, covariance is the mean value of the product of the deviations of two *variates*, i.e. the set of all random variables following a given probabilistic law from their respective means. Covariance determines the extent to which two random variables X and Y covary, i.e. change together in the same way (Pourahmadi 2013).

However, it is difficult to compare covariances among data sets having different scales. For instance, a value representing a strong relationship for one data set might represent a very weak one in another. The correlation coefficient overcomes this limitation by normalising the covariance as the product of the standard deviations of the variables and creating a dimensionless quantity suitable for the comparison of data sets having different scales (Shevlyakov and Oja 2016).

Correlation can therefore be considered the scaled version, or *standardised* form, of covariance.

Correlation measures the relationship between the varying of two entities, e.g. signals or waves, through correlation coefficients evaluating *similarity*.

More precisely, the correlation between X and Y is the covariance of the corresponding *standard scores*, where in statistics the standard score is the signed number of standard deviations by which an observation or data lies above the mean (Urdan 2016).

In signal processing, the *cross-correlation function* quantifies, for instance, how well correlated two waves are. More precisely, the cross-correlation is a measure of similarity between two series as a function of the time lag of one relative to the other, i.e. when the time lag is accounted for, the fit improves.

In physics, *autocorrelation* (Broersen 2010) of a signal takes place when a value adopted at a given time correlates with another at a different time, i.e. the correlation of the signal with itself at different points in time. This allows one to reconstruct or anticipate values over time.

Within a population of interacting entities, one can consider the *correlation length* (or correlation radius), i.e. the set of correlated elements, namely the spatial span of the correlation. When the spatial span is as large as that of the entire population, i.e. the number in the set of the correlated elements is equal to the total number of elements, then we have *scale-free correlations* (Cavagna et al. 2010; Hemelrijk and Hildenbrandt 2015; Altamura et al. 2012).

We also recall that analogy (Gilboa et al. 2015) is a *partial* relationship of similarity relating to only some of the variables necessary for representing a phenomenon. For example, if the intensities of the same variable representing two phenomena vary in a proportional way but their frequencies do not.

Several statistical approaches are available to identify correlations, such as Multivariate Data Analysis (MDA) (see Hair and Black 2013), Cluster Analysis (Everitt and Landau 2011), Principal Component Analysis (PCA) (Jolliffe 2002), Recurrence Plot Analysis (RPA) (Webber et al. 2016), and Recurrence Quantification Analysis (RQA) (Webber and Marwan 2016).

Synchronisation is a correlation occurring when changes over time are regularly iterated.

4.3 Coherence

There are several ways to understand coherence.

In philosophy, for instance, one meaning relates to the *lack of contradictions*. In this way, *coherent reasoning* is free from contradictions.

This assumes linear approaches, as opposed to, for example, Hegelian *dialectics,* when considering development through the stages of thesis, antithesis, and synthesis (Rosen 1982).

On the other hand, dialectics relates to intellectual exchanges of ideas allowing *temporary* contradictions for eliciting truth.

We consider now the relationships between correlation and coherence.

Synchronisation is the simplest coherence, where the way of changing is iterated in the same way.

Technically, in signal theory, coherence is similar to correlation. When two signals perfectly correspond to each other at a given frequency, the *magnitude of coherence* is *1*. If they do not correspond at all, coherence is *0*.

As stated above, correlation measures the *similarity in the ways of changing* of two variables, such as the prices of two different consumer products.

This may be considered as *microscopic correlation*.

However, when dealing with complex systems, coherence relates to the maintaining of the same collective property, e.g. patterns or density, by *structurally variable* interacting collective entities (Minati and Pessa 2018, pp. 65–69). This is the case for oscillators or boids having multiple different interactions per instant, while maintaining similar behavioural patterns over time.

We have *diffused correlation, understandable simply as coherence*, in the case of long-range, scale-free correlation, and *polarisation,* as introduced below. In the case of scale-free correlations, the *correlation length* coincides with the total extension of the systems and they coincide with coherence of the entire population of individual entities (Cavagna et al. 2010; Stanley et al. 2000). *In this case,* coherence *can be thought of as* global dynamical ordering (see Sect. 4.4).

Structural Dynamics

As stated above, the purpose of this contribution is to introduce approaches for possible conceptual understanding of the ways in which diffused correlation, i.e. coherence, occurs.

With this purpose in mind one can distinguish between:

- The *dynamics* of collective entities interacting through the *same* rules of interaction, in various cases occurring at different times, with different durations or intensities, with variable parameters, and in different combinations.
- The *structural dynamics* of collective entities interacting through *different* rules of interaction over time. In this case, we deal with *structural changes* when, for different reasons, not only parameters but also rules can change. This may occur, for instance, because of the occurrence of sequences of phase transitions that change the *nature* of the systems of collective entities, of sequences of different processes of self-organisation (Minati 2016a; Minati and Licata 2012), or when reaching criticalities "… living systems as 'coherent structures' in a continual (extended) critical transition. The permanent state of transition is maintained, at each level of organization, by the integration/regulation activities of the organism, that is by its global coherent structure. … However, the 'coherent critical structures' which are the main focus of our work cannot be reduced to existing physical approaches, since phase transitions, in physics, are treated as 'singular events' … Thus, a living object is understood not only as a dynamic or a process, in the various possible senses analyzed by physical theories, but it is a *permanent critical transition*: it is always going through changes…" (Bailly and Longo 2011, p. 18). Yet we may consider the cases of an evolutionary process leading to transitions (Giuliani and Zbilut 2008), such as through learning or the passing of energy thresholds. This may be intended as *dynamical order* (Manrubia and Mikhailov 2004).

The coherence of the collective behaviour of populations of interacting entities pertains to *compatible*, i.e. non-contradictory, physically possible, sequences of new configurations maintaining the same collective properties over time, including evolutionary patterns, scale invariance, and long-range correlations (see Sect. 5).

This applies to emergent properties that are *robust* to changes and maintain emergence in sequences of new configurations, that is, in several *equivalent ways* selected from among all those properties that are compatible and available. The robustness of emergent properties is related to the coherence of the sequences of new configurations, where the compatibility of the sequences is a necessary condition for coherence (Minati and Licata 2013).

As an example useful for consolidating these ideas, we may consider the coherence of the collective behaviour of a swarm of mosquitoes around a centre, usually a light, or seagulls circling around a pile of garbage. In these cases, the coherence of orbits should be considered due to their compatibilities and equivalences generated in turn by the respect of rules such as those in the model introduced by Reynolds (1987) in computer graphics to simulate the collective behaviours of flocks, herds, and schools. Such microscopic rules state that the positional choices (the ways for this emergence to arise) for each interacting agent must maintain distances not exceeding a suitable maximum and minimum and are made in such a way as to ensure:

- Alignment, given by pointing towards the average direction of the local adjacent agents
- Cohesion, given by pointing towards the average position of the local or adjacent agents

while boids are all considered able to properly vary speed, direction, and altitude.

Another approach, as mentioned in Sect. 2, is related to considering topological rather than metrical distance (Ballerini et al. 2007).

The previous approaches possess a microscopic nature suitable for simulations, i.e. they are easy to be prescribed.

More generally, coherence may be considered, for instance, as a result of diffuse correlation, scale invariance. and polarisation (see below). However, there are different *ways* (assumed here to be compatible, equivalent. and interplaying) by which interacting elements may keep such properties. We are interested here in exploring possible conceptual representations of such ways.

We mention in the following section the mesoscopic representation where clustering represents families of behaviours considered equivalent. For instance, clusters as groups respecting the same thresholds, e.g. ranges of minimum and maximum values (see section "Meta-Structural Properties"), or being ergodic, or algorithmically determined, e.g. minimising the total intra-cluster variance. Another approach based on Networks is considered in section "Network Properties".

Approaches

In the following, we deal with approaches having a mesoscopic nature (Giuliani 2014), based on clusters, which are suitable for considering the interplay between equivalences as introduced in Sect. 2. In complex systems, the coherence of the ways of behavioural changing through configurations and patterns, such as maintaining the *status* of flock, swarm, market, or by keeping acquired properties through dissipative structures, can be suitably modelled and represented by using clusters. For example, we may consider the two following possible cases: ergodic and computational clustering.

The first case occurs when systems are ergodic (Minati and Pessa 2006, pp. 291–320). The concept is clearly conveyed by considering a population of entities interacting in such a way that *if x% of the population is in a particular state at any moment in time, we can assume that each entity of the population spends x% of time in that state*. Entities take on the same roles at different times and different roles at the same time, but with the *same* percentages. This relates to the conceptual interchangeability of entities playing the same roles at different times. *Percentages may be considered as* equal *at a suitable threshold*.

In this case, the statistical properties can be deduced from a single but sufficiently long, random sample.

This opens up the possibility of considering different clusters composed of entities connected by *different ergodicities* when considering different simultaneous states. More realistically, we may consider quasi-ergodic systems where the degree of ergodicity is variable, but involves sufficiently high percentages of entities of the population under study. Ergodicity (Coudène 2016) is applied in various disciplines such as economics (Domowitz and El-Gamal 1997, 1999; Landon-Lane and Quinn 2000), geology (Paine 1985), and sociology (Barbi et al. 2004; Lee 1974, 1978, 1985, McCaa 2001).

Therefore, we are talking about possible ergodicity or quasi-ergodicity among clusters, representing sequences of new configurations, allowing their coherence (Minati et al. 2013).

In this case correlation is given by ergodicity.

The second case is based on clustering techniques. We consider clusters consisting of different groupings of entities, when entities belonging to the same cluster have the same covariance, being equally correlated. Each cluster corresponds to different local correlation lengths.

Such clustering is possible by using suitable available techniques such as k-means (Wu 2012) *where the objective is to minimise the total intra-cluster variance.*

At this point with an available population of clusters and their properties per instant, such as their number of components (and their possible multiple membership in clusters), it is possible to consider properties of *correlations among clusters*, e.g. among their numbers of components, statistical aspects, and validity of the same thresholds, as for Meta-Structures (see section "Meta-Structural Properties").

In this case, the concept of coherence is related to correlations between the ways of changing of *clusters of components instead of the ways of changing of single components*.

Within a population of interacting entities, global coherence can be considered as being given by the correlation among clusters.

Correlations among clusters may be studied using various suitable approaches such as the statistical techniques mentioned above.

Shortly, we discuss how incoherence can be understood as *incompatibility*, in the sense introduced above (see Sect. 2.1), between sequences of configurations.

Moreover, incoherence may be understood as being due to the loss of emergent properties over time, such as a loss of consistency of the status of flocks veering towards disaggregation or the loss of scale invariance.

The issue of incoherence as incompatibility between sequences of configurations may not apply within the conceptual context of *multiplicity*. For example, Multiple Systems or Collective Beings occur when the coherence of the global system is given, for instance, by ergodicity of entities simultaneously or sequentially interacting in various ways and playing interchangeable, multiple, or overlapping roles (Minati and Pessa 2006, pp. 97–134, see Sect. 5.1). In this case, compatibility is not necessary for *all* systems of a Multiple System such as the Internet or global markets.

However, to avoid an over-simplified understanding of a complex system, the acquired emergent properties must be considered not necessarily as being easily distinguishable. In a complex system, properties can be composite, interfering or having different independent life spans. This corresponds to multiple, temporal, partial, composite, and duration-varying correlations.

An example is given by detecting the *persistence* of the same flock while groups of constituent boids form temporary differently *correlated* clusters, where such clusters might be comprised of those boids with the same instantaneous velocity, distance, direction, or altitude.

Using an analytical approach, the problem of representing coherence of such multiplicity is almost intractable at a microscopic level, where each case should be hypothetically formalised in the same way, if not limited to considering different values of a single parameter.

Such coherence may be represented by properties such as power laws, network properties, scale invariance, or meta-structural properties (see Sect. 5).

4.4 Polarisation and Global Ordering

Within a population of interacting entities, e.g. swarms or flocks, it is possible to consider the degree of *global ordering* measured, for instance, by *polarisation* (Cloude 2014). In physics, polarisation relates to phenomena such as waves in liquids or gases only oscillating in the direction of propagation of the wave, or to light *vibrating* mostly in one direction.

Within a population of interacting entities,

- There are instantaneous, differently polarised clusters consisting of differently correlated and possibly dispersed entities, having, for example, the *same* direction.
- Polarised clusters correspond to the correlation lengths, i.e. the extent of the correlation. When the extent coincides with the entire collective system, we have scale invariance and the population is all polarised.

Such correlations may be partial (i.e. some entities may be not involved), temporary, unrelated, equally iterated, unstable, or *all correlated* (Cavagna et al. 2010; Hemelrijk and Hildenbrandt 2015).

In the case of multiple polarisations, their possible correlations establish the coherence of collective interacting entities.

We can state that there are variable clusters of local, temporal, correlated elements which are, however, all correlated as clusters.

Polarisation is a particular example of the second case introduced above in section "Approaches", as it applies to polarised clusters.

The concept of global order also relates to robustness, in the sense of being able to recover from temporary disorder due, for example, to very local inconsistencies that are occasional and disperse. Local inconsistencies should be intended as quasi-correlations, quasi-covariances of polarised clusters, and quasi-polarised clusters when correlations may be partial but sufficient to maintain global coherence and consistency of the collective behaviours (Giuliani and Zbilut 2008; Manrubia and Mikhailov 2004).

5 The Coherence of Emergence in Complex Systems

In the first part of this Section, we briefly list some examples of phenomena having structural dynamics which can be considered consistent with, if not fully possessing, the interplay between compatibilities and equivalences which facilitates or induces processes of emergence.

In the second part, we list some properties of processes assumed to occur in various ways representing such interplay. Some properties are related to the classical Dynamical Systems Theory (Alligood and Yorke 2009) where we consider a generic autonomous system of ordinary differential equations

$$dQ_i / dt = f_i\left(Q_1,\ldots,Q_n\right), \quad i = 1,\ldots,n$$

where:

- time, t, is the only independent variable.
- Q_i *is* the ith dependent variable.
- the total number n of the latter gives the number of *degrees of freedom* of the system.
- The symbols f_i denote functions of their arguments.

The classic pre-complexity (Minati 2016b) so-called *General System Theory* (Von Bertalanffy 1968) introduced by the mathematical biologist Ludwig von Bertalanffy (1901–1972), considered a system as being represented by

- Suitable state variables Q_1, Q_2, …, Q_n, whose instantaneous values specify the state of the system
- A system of *ordinary differential equations*, such as:

$$\begin{cases} \dfrac{dQ_1}{dt} = f_1(Q_1, Q_2, \ldots, Q_n) \\ \dfrac{dQ_2}{dt} = f_2(Q_1, Q_2, \ldots, Q_n) \\ \ldots\ldots\ldots\ldots\ldots\ldots\ldots\ldots \\ \dfrac{dQ_n}{dt} = f_n(Q_1, Q_2, \ldots, Q_n) \end{cases}$$

This system is intended to specify *how* the change in value of a given state variable, affects all other state variables representing *interactions* intended to occur when *the behaviour of one affects that of the other.*

However, other properties of complex phenomena having dynamical equivalences, multiple coherences, and dynamical temporary incoherencies (as introduced in Sect. 2) are analytical intractable, i.e. they cannot be *zipped* nor exhaustively represented with analytical formulae as for the Dynamical Systems Theory. In such cases, it is possible to use *non-ideal models* (mixtures of general principles and of specific choices requiring computer simulations) based, for instance, on neural networks or cellular automata, in contrast with *ideal models* (Minati and Pessa 2006, pp. 161–189) having a top-down structure, based on general principles assumed to be universally valid, i.e. covering the widest possible spectrum of phenomena, such as Synergetics (Haken 1983), Spontaneous Symmetry Breaking and Noise-induced phase transitions allowing the deduction of particular consequences and forecasts.

Otherwise to deal with high levels of dynamical complexity, it is possible to consider approaches based on Meta-structures (see section "Meta-Structural Properties") or Networks (see section "Network Properties").

Specifically, even if a precise distinction is not possible, in the first part of this Section we list *phenomena almost compatible with processes of emergence, suitable to host processes of emergence, and able to establish themselves as processes of emergence.* We consider as examples processes of establishment of attractors, bifurcation points, chaotic behaviour, dissipation, domains of coherence, multiple synchronisations, and Multiple Systems.

Also in the second part, we list some properties of representations characterising the evolutionary paths of the phenomena considered above and of their multiple observable processes. Their multiple occurrences within a framework of non-decidability; validity of the uncertainty, complementarity principles; and incompleteness, for instance, in a DYSAM-like way (see Sect. 2.3), corresponds to the dynamical multiplicity of complexity. These properties and processes are intended as signs, clues, *theory-substitutive* concordances and correspondences, and possible trademarks of the coherence of interplaying compatibilities and equivalences establishing processes of emergence (Schroeder 2009). This allows *scale-free* detection of processes of emergence, when emergent properties occur on a phenomenological scale other than that of constituent entities. For instance, collective behaviours or entities at great distances, e.g. boids flying at relative distances making flocks undetectable to the naked eye. Such properties are characteristics typical of complexity;

featuring processes of emergence such as establishing attractors in models, meta-structural properties, network properties, and non-equivalence; following power laws, scale invariance, and symmetry breaking; and having unpredictability. Finally, we consider the constructivist role of the observer.

As stated above, the dynamics of such kinds of complex systems are known only *a posteriori* and the idea *to zip* the essential characteristics of change into a set of ideal equations, typically a Lagrangian or Hamiltonian formulation based on general symmetry or conservation principles, is unsuitable. As power laws and scale-freeness are *clues* of complexity, i.e. the occurrence of processes of emergence and self-organisation, properties of the behaviour of systems selecting from among *equivalent* possibilities, for instance, respecting the degrees of freedom, may *profile* complex behaviours.

5.1 Examples of Phenomena Compatible with Emergence, Suitable for Hosting Processes of Emergence, and Possibly Establishing Themselves as Processes of Emergence

We consider here the establishment of:

Attractors (Phenomenon)

We focus here on *real attractors*, i.e. physical phenomena, as in fluid dynamics, e.g. whirlpools, and having the nature of vortices such as for swarms of mosquitoes flying around a light or seagulls circling around a pile of garbage. In the first case, the attractor at the centre is of an energetic nature while in the second case is of a cognitive nature.

In the case of multiple lights, piles, or dissemination, flight activities will vary correspondingly (see section "Attractors (Property of Models)" for a more general discussion).

Chaotic Behaviour

We may distinguish between *deterministic chaos* (Lorenz 1963; Sparrow 2013) and *stochastic chaos* (Freeman et al. 2001).

Deterministic chaos can be identified with an apparently random motion stemming from deterministic equations and can be associated with time behaviours characterised by:

- Long-term unpredictability
- High sensitivity to initial conditions

The first property (long-term unpredictability) is typical of even non-deterministically chaotic behaviours, such as *noise*.

The second property is typical only of deterministic chaotic behaviours.

A celebrated example is the climate (Goosse 2015) which represents the essence of deterministic chaos. In this case, at any point in time, the difference between two behaviours associated with two different initial conditions grows exponentially with time *however small* their difference may be.

The Lorenz attractor, a solution to the Lorenz equations, displays remarkable behaviour and landmarks in the field of deterministic Chaos. The model was intended to simulate medium-scale atmospheric convection (Lorenz 1963; Sparrow 2013).

By plotting the behaviour of its numerical solution, we obtain a structure which weaves in and out of itself around two attractors, denoting the so-called *butterfly effect*.

Stochastic chaos occurs within systems endowed with noise both from inner sources and from the external environment.

While deterministic chaos is characterised by a low dimensionality of attractors and by the fact that time is the only independent variable, stochastic chaos is characterised by having attractors of high dimensionality, the presence of noise-created and noise-sustained structures, spatio-temporal unpredictability, and spatially independent variables.

Stochastic chaos can be observed mainly through computer simulations as well as experiments.

Dissipation

The term *dissipative structures* established by processes of dissipation (Minati and Pessa 2006, pp. 171–177, see below) was introduced by Ilya Prigogine (Prigogine 1967, 1981, 1998) referring to situations of *coexistence of change and stability*. In these situations, suitable processes, for example, metabolic, keep such systems far from thermodynamic equilibrium, i.e. thermodynamic death.

The attribute *dissipative* refers to systems where energy dissipation in non-equilibrium conditions allows the emergence of ordered structures. The stability of dissipative structures is due to their ability to transfer a large amount of entropy to their environment, rather than due to the classic understanding of low entropy production. Entropy is intended as an index characterising the microscopic disorder of the system under study. The entropy growth indicates a trend towards a more disordered phase, such as from solid to liquid or gas.

Such systems, far from thermodynamic equilibrium, are able to dissipate the heat generated to support themselves, leading to the emergence of ordered configurations. In other words, this allows for processes of self-organisation.

Furthermore, such systems contain continuously fluctuating subsystems giving rise to processes corresponding to *bifurcation* points (Stein 1980). It is therefore impossible to predict whether dissipative systems will degenerate into a chaotic situation or reach a higher structural, organisational level. In the latter case, *dissipative structures* are established. The attribute *dissipative* is related to the fact that they need incessant feeding of matter/energy to continue their existence and that their existence is limited by their related ability to dissipate matter/heat.

We may say that a dissipative structure arises as a balance between the dissipation, diffusion, and the nonlinearity enhancing its inner fluctuations.

An example of a dissipative non-living structure is a vortex in a flux of running water. In this case, water continuously flows through the vortex but its characteristic funnel shape shrinking into the spiral remains. There are similar structures in atmospheric phenomena such as hurricanes.

Analogously, a living dissipative structure requires a constant flow of matter, such as air, food, water, and, in certain cases, light.

Domains of Coherence

The most often considered phenomenon consisting of the establishment of domains of coherence is the physics of the emergence of quantum coherence in water (Hirst 2013).

In the classical view considered here, the phenomenon concerns the establishment of domains (e.g. subsets, multiple systems, and clusters) within the same population of entities, where these domains have different possible coherences given, for instance, by different, localised lengths of scale invariance.

The concept of domain of coherence may be considered to coincide with that of correlation length. In this case, each domain has its own specific correlation. As for Multiple Systems (see section "Multiple Systems"), we may hypothesise the possible occurrence of multiple domains of coherence.

Multiple and Remote Synchronisations

Consider, for instance, the case of multiple, correlated oscillators or clusters establishing coherence and domains of coherence when correlation is given by synchronisation. Multiplicity may relate to different oscillators and clusters either identically or differently synchronised, e.g. through local or possibly spatially dispersed communities. An interesting case occurs as remote synchronisation when two oscillators not directly linked but both connected to a third unit, can become synchronised even if the third oscillator does not synchronise with them. This effect relies on the modulation performed by the intermediary node (the third unit), which allows for the passage of information between two otherwise unlinked neighbours (Gambuzza et al. 2013). Such a phenomenon leads to heightened synchronisation between distant segments in an experimental ring (Minati 2015).

Multiple Systems

In Multiple Systems, the same components can play interchangeable, multiple, and overlapping roles. This occurs, for instance, when a specific action, spatial position, or value of a variable has different meanings or effects depending on the corresponding systems they establish. Interchangeable, multiple, and overlapping roles may occur when components simultaneously or sequentially interact in different ways, dynamically constituting sequences of different systems. Examples of Multiple Systems include electronic devices where values adopted by the same components have simultaneous multiple meanings, e.g. regulatory and related to safety, and programmes and nodes of the Internet simultaneously allowing user profiling and geolocation.

Multiple Systems are called Collective Beings when the constituent agents are autonomous systems, i.e. possessing cognitive systems sufficiently complex to perform inferences, have and process memory, and form representations allowing them to decide their behaviour and mode of interaction. Examples of Collective Beings include flocks where the same boid is simultaneously a neighbour of others and also plays an intermediated role for information transfer. “Correlation is the expression of an indirect information transfer mediated by the direct interaction between the individuals: Two animals which are outside their range of direct interaction (be it visual, acoustic, hydrodynamic, or any other) may still be correlated if information is transferred from one to another through the intermediate interacting animals” (Cavagna et al. 2010, p. 11865). Another example is given by individuals who can simultaneously act as members of families, corporate systems, road traffic, markets, and telephone user networks.

Another issue relates to community detections in complex systems and networks (see, for instance, Missaoui and Sarr 2015, and Sect. 5.2).

5.2 *Examples of Properties Intended as Signs, Clues, Theory-Substitutive Concordances and Correspondences, and Possible Trademarks of Emergence*

We consider here properties such as:

Attractors (Property of Models)

Contrary to the usual geometrical space, in physics *the phase space* is an abstract space where suitable state variables $Q_1, Q_2, \ldots, Q_n$ of the system are associated with a coordinate axis. It is possible to represent in a graphical way such n-dimensional space only in the particular cases of dimensions ≤ 3.

The entire system behaviour can be represented as the motion of a point along a trajectory within this space.

In the case where the system approaches a given domain of this space as time tends to *infinity*, that domain is called an *attractor* of the system dynamics.

It is possible to consider three kinds of attractors:

- *Fixed points*, consisting of isolated equilibrium states
- *Periodic attractors*, typical of systems exhibiting periodic oscillations in the long term
- *Strange attractors*, corresponding to fractal domains of phase space, usually present in chaotic systems

Attractors provide information about the *kind* of system behaviour.

The analysis of non-linear systems considers the topological characteristics of their attractors and is known as *qualitative analysis*.

In short, an attractor may consist of either a single point, a finite set of points, a curve, or a manifold around which, starting from any variety of initial conditions, the evolutionary path of a dynamical system tends to evolve.

We consider here attractors related to models representing system behaviour when the subsequent time values representing the steps or points of the evolutionary path belong or quasi-belong to a graph around the attractor, i.e. respecting its basin of attraction. Evolutionary paths around an attractor are robust to perturbations (Ruelle 1989).

The Lorenz butterfly attractor is a celebrated example (Lorenz 1963; Sparrow 2013), developed to model atmospheric conditions and then used in interdisciplinary studies to model chaotic phenomena. See section "Chaotic Behaviour".

Bifurcation Points

This term denotes a change in the topological structure of the system and the number or type of attractors as a consequence of small, smooth changes in parameter values.

In simple cases dealing with a single parameter, a bifurcation occurs when the value of a parameter (named *bifurcation* or *critical parameter*), crosses a *critical value*.

This critical value plays the role of separator between the structures, with one structure having values of the bifurcation parameter less than the critical value, and the other having values greater than the critical value. In general, a bifurcation is associated with another phenomenon known as *symmetry breaking*, see section "Symmetry Breaking".

From this point of view bifurcation phenomena are suitable to classically describe self-organising systems and phase transitions in physical systems (Cicogna 1981; Aihara et al. 2015).

Furthermore, one can distinguish between three different classes of bifurcations:

- *Subtle bifurcations*, in which the *number* of attractors remains constant, but their types change
- *Catastrophic bifurcations*, in which the number of attractors change
- *Plosive bifurcations*, in which the attractor size undergoes a sudden, discontinuous change as the bifurcation parameter crosses a critical value

However, *Bifurcation Theory* (Kuznetsov 2004) is unable to give a full explanation of phenomena of emergence (Nitzan and Ortoleva 1980).

Meta-Structural Properties

The term meta-structure relates to the instantaneous multiple structures of interactions establishing coherent collective behaviours. Such interactions may have different temporal durations, different initial and final moments, different intensities, and may superimpose or combine in any way. They constitute the dynamical structure of collective behaviours. Because of their multiplicity and variability as mentioned above they are named “meta”.

Meta-structural properties are intended to model dynamics and coherence of such meta-structures. Meta-structures are analytically intractable. We considered approaches based on properties of instantaneous multiple clusters that are suitable to represent the structural dynamics mentioned above (Minati and Licata 2012, 2013, 2015; Minati and Pessa 2018, pp. 102–116); Minati et al. 2013; Pessa 2012).

Meta-structural properties are typically inter-cluster properties (see Sect. 4.3) and include:

- The correlation between the number of agents constituting clusters over time, such as agents clustered by speed, altitude, or direction
- Properties, regularities, and the distribution of the number of agents constituting clusters and, conversely, of agents not belonging to any cluster over time

Network Properties

The Science of Networks (Baker 2013; Barabási 2002; Dorogovtsev et al. 2008; Estrada 2016; Lewis 2009; Motter and Albert 2012; Newman et al. 2006; Newman 2010; Valente 2012) represents systems as networks and systemic properties as network properties. Particular properties of complex systems may be represented as properties of networks.

For instance, complex systems may be considered as being represented by complex networks (Cohen and Havlin 2010) having properties such as being:

- Scale-free, which occurs when the network has a high number of nodes with few links or a small number of nodes (hubs) with a high number of links. In scale-free networks, the probability that a node selected at random will possess a particular number of links follows a power law (see section “Power Laws”).

The property of a network being scale-free strongly correlates with its robustness to failure by establishing fault tolerant behaviours. Examples include the Internet and social collaborative networks.

- Small-world, occurring when most nodes are not close neighbours, but most nodes can be reached from every other node via a small number of intermediate links. This property is also considered to increase robustness. Examples include electric power networks and networks of brain neurons.

Furthermore, there are other network properties to be considered such as:

- Cluster coefficient.
  The cluster coefficient is an important measure of the network structure of nodes that are close to each other, or specifically the network cohesiveness. In many complex networks, it is possible to find clusters that are subsets of the network and possess a high level of inner connectivity. In this regard, the clustering coefficient measures the degree of clustering of the node's neighbourhood. In particular, the Cluster coefficient is intended as a measure of the likelihood that any two nodes possessing a common neighbour are themselves connected. An example is given by social networks of friends who generally know each other.
- Degree distribution.
  The degree of a node is the number of neighbours of the node. The degree distribution is the probability distribution of the node degrees over the entire network.
- Fitness.
  The way the links between nodes change over time depends on the ability of nodes to attract links.
- Idempotence.
  We mention also the property of self-similarity of links and paths. This property is linked to a form of matrix idempotence. A generic matrix M is said to be idempotent if $M = M^2$.

Consider a matrix representing, for instance, a graph G. Idempotence of this matrix highlights self-similarity of the forms of network suitable to represent pattern formation and network dynamics. More realistically in practice, the concept of quasi-idempotence is considered, where $M \approx c + kM^2$ (Minati et al. 2017), where c, k are suitable constants.

Non-Equivalence

We refer to equivalence and non-equivalence, and their balancing in collective behaviours, as mentioned in Sect. 2 concerning admissible and compatible changes.

We consider here how non-equivalence does not presuppose or mean *incompatibility*. The point is that there are *several non-equivalent ways to maintain coherence*. For instance, we may have variable correlations among temporal, local, and different locally non-equivalently correlated communities established by agents of a collective behaviour and not only having the same long-range correlation.

As a matter of fact, within processes of emergence the compatibility of sequences of non-equivalent properties does not *prescribe* the same coherence, but rather represents how, at any step, several possibilities are available for the evolution of the process.

The resulting configuration, i.e. the collective behaviour, consists of non-equivalent but compatible and correlated steps, e.g. sequences of non-equivalent, but coherent patterns.

This point relates to multiple domains of coherence, multiple and remote synchronisations, and Multiple Systems as mentioned in sections "Domains of Coherence", "Multiple and Remote Synchronisations", and "Multiple Systems".

Power Laws

A power law or scaling law is the form taken by a remarkable kind of functional relationship between two variables X and Y, such as $Y = kX^{\alpha}$, where α is the power law exponent and k is a constant. In this case, one quantity varies as a power of the other one.

The particularity of the distribution allowed by power laws concerns the suitability to represent concentrated aggregation of high values and much diluted, so-called *long tail*, low values, of Y.

When the frequency of an event varies as a power of some attribute of that event, e.g. its size, the frequency is said to follow a power law (Schroeder 2009), such as for size vs. number of corporations, levels of wealth vs. number of people considered, magnitude vs. number of earthquakes, size of cities vs. sizes of population, and word frequencies (relatively few are commonly used).

Considering the distribution of income Y, a high level of income relates to only a few people, while low-to-medium incomes relate to a large majority of people.

In biology, we may consider the case studied by allometry, concerning the relative growth of an organ or part of an organism compared to the whole body, or between the size of the body compared to its metabolic rate (energy consumed by an individual during a unit of time). Where M represents the animal mass, it has been found that the metabolic rate $R = M^{3/4}$ for *any* kind of animal (Brown and West 2000).

Furthermore, power laws are scale invariant.

Scale Invariance and Self-Similarity

Scale invariance (Henriksen 2015) is the feature of entities not changing their properties, e.g. geometrical properties of patterns in morphology, regardless of a change in dimensions, for instance, of resizing, or number of components.

A related technical term for this transformation is *dilatation* (also known as dilation). Dilatations can also form part of a larger conformal symmetry. We can imagine the property of scale invariance as a *continuous homogeneous process of positive or negative dilatation*.

Another related concept is that of *self-similarity*, where a property of an object is, at any time of its evolution, similar to a part of itself. This is the typical case of fractals (Bunde and Havlin 2012; Falconer 2013). The first example of a fractal was introduced by Helge von Koch (1870–1924) elaborating some previous ideas considered by the mathematician George Cantor (1845–1918) with the so-called *Koch curve*. The mathematician Benoit Mandelbrot (1924–2010) introduced the specific term fractal. Examples in nature are snowflakes, ramifications of a tree, leaves, and flower structures.

Scale invariance is a form of self-similarity where at any positive or negative dilatation there is a larger or smaller piece of the object that is similar to the whole.

Symmetry Breaking

In general, a bifurcation (see section "Bifurcation Points") is associated with the phenomenon known as *symmetry breaking*.

The expression *symmetry transformation* denotes a transformation of suitable variables in the evolution equations of a given system. From a mathematical point of view, the solutions of dynamical evolution equations are invariant in form with respect to symmetry transformations, e.g. rotation. These rotations are only a proper subset of symmetry transformations leaving invariant the form of the evolution equations themselves.

However, this transformation can act both upon the form of these equations, as well as upon the form of their solutions.

We have symmetry breaking when a symmetry transformation leaves the form of the evolution equations invariant, but changes the form of their solutions.

A typical example is given by considering matter which, at a given temperature, is paramagnetic.

The form of the equations describing the motion of constituent atoms is invariant with respect to particular symmetry transformations consisting of space rotations around a given axis. The solutions to these equations also have the same invariance. However, if the matter is exposed to an external magnetic field, whatever its direction, this will give rise within the material to an induced field aligned with the external one. When the temperature is decreased, there is a critical point, named the *Curie point*, where the transition from a paramagnetic to a ferromagnetic phase occurs. This gives rise to an internal macroscopic magnetic field. The presence of such a field leads to the existence of a preferred direction of alignment for the atoms, i.e. that of the internal magnetic field. Even if the form of the equations describing the motions of the atoms does not cease to be invariant with respect to the symmetry transformations constituted by spatial rotations, their solutions do not, because the preferred direction breaks such invariance (Mitra 2014).

The related concept of *spontaneous symmetry breaking* relates both to classical and quantum physics (Strocchi 2010).

Unpredictability

We consider here the unpredictability of emergent behaviours as given step by step by *decisions* between equivalences and tolerable temporary non-equivalences. This facilitates a robustness of the process (see Sect. 2). Such decisions occur, for instance, due to the instantaneous predominance of fluctuations and noise, and the occurrence of criticalities such as bifurcations and symmetry breaking crossing a critical point. These events decide the sequence of the process of emergence (see above). Such decisions are undecidable and unpredictable in the sense that no procedures, no algorithms are available to decide: decisions are conceptually non-automatable due to equivalences (see Sect. 2.2). The collective behaviour can occur in different, although equivalent, ways.

The Constructivist Role of the Observer

Ontology is intended as *represented knowledge* (Jakus et al. 2013).

The changes in ontologies are of interest as being relevant for representing structural changes in a system and related levels of coherence during processes of emergence. Such structural changes include acquisition or loss, and changes in properties.

Regarding the presence and evolution of levels within systems we may consider, for instance, Baas (1994) and Heard (2006).

We underline how the constructivism of the observer must absolutely not be reduced to relativism, e.g. arbitrariness of points of views and ordering per arbitrary importance (Fields 2016).

This point relates to the abductive (Gabbay and Woods 2005) selection and invention of variables and models to effectively represent phenomena in which the observer is involved, both as a passive and an active subject, e.g. designer or manipulator (Licata 2008; 2012; Steffe and Thompson 2010). It is critical to invent the models and determine the degrees of freedom to usefully represent the system under study for the benefit of the observer.

We conclude this Section by stressing how phenomena and properties of processes of emergence, some of which are mentioned above, may occur in different dynamical combinations, superimpositions, and varieties to be further studied within the conceptual context of theoretical incompleteness and equivalence.

6 Simulations

Here, we discuss how the dynamical coherence of the interplay and combinations of phenomena and processes discussed above may be simulated. Models used for simulations in this case to partially represent coherence occur for completely analytically defined processes having random variables (Minati and Pessa 2018, 204–205).

In this chapter, we tentatively describe the multiplicity of the natures (See Sects. 5.1 and 5.2) of processes, phenomena, and properties which allow and theoretically (incompletely) constitute emergence. We attempt to identify such properties and phenomena without using the reductionist assumption that precise separation and *disassembling* of their interplay constituting emergence is possible. The purpose is to tentatively identify such multiple natures and their interplay allowing a general conceptual framework to deal with processes and phenomena of emergence which are theoretically incomplete. Examples of specific purposes are to:

- Identify possible dominant aspects allowing *focused* models
- Identify aspects suitable for effective modifying interventions, i.e. those able to orient, vary, sustain, or even disintegrate the process. Examples include environmental interventions and perturbations
- Avoid any reductionist understanding of emergent properties and avoid corresponding reductionist approaches such as assuming it possible to identify the optimum model, completeness, context independence, and stable structural dynamics

However, such irreducible multiplicity may be simplified and still be sufficient for finalised representations such as simulations. Simulations may be based on assuming the extensive validity of simplifications of the interplay discussed above and assuming analytically tractable interactions *as significant or even predominant.* This is possible, for instance, by assuming the dominance of one specific property such as the occurrence of chaotic behaviour, power laws, or scale invariance.

Consider the simplification consisting of assuming dominance of one or only several aspects, disregarding, for instance, other aspects such as theoretical incompleteness and unpredictability due to undecidability. In simulations, the ignored, disregarded parts can be instead represented by randomness, acceptably equivalent to their *cumulative resulting effect.*

Many examples of flock simulators have been introduced following the approach introduced by Reynolds (1987, http://www.red3d.com/cwr/boids/) in computer graphics, such as in https://sourceforge.net/projects/msp3dfbsimulator/?source=directory. An interesting overview is available (Vicsek and Zafeiris 2012).

The issue of simulation of collective behaviours (Rossetti 2015) is usually performed by agent-based approaches (Taylor 2014) relating, for instance, to crowd behaviour, emergency evacuation, markets, and vehicular traffic. However, the repeatability of simulations should not be confused with achievement of final representations, but as sources of clues and hypotheses, generators for complex phenomena in the absence of robust theorisations able to explain.

We mention how the content of this chapter also concerns the issue of the so-called *big data* relating to the availability of enormous quantities of data *without knowing what to do with them* (see Calude and Longo 2016). *Enormous quantities of data are available, but we did not* decide *to collect them according to pre-established models or theories, e.g. prices, lengths of communications, and tickets.* Some of the approaches mentioned above may be considered for *profiling*, i.e. finding meaningful dominances and correspondences within Big Data available for *phenomenological*, for instance business, and not theoretical or modelling purposes.

7 Conclusions

We presented possible analytical, theoretically incomplete understandings of the phenomenological occurrence of processes of emergence. We particularly considered processes of emergence of collective behaviours taking place in physical *3D* space because of their suitability to represent tractable cases, generalisable to non-physical spaces such as spaces of prices.

The main purpose of this chapter is to contrast the misunderstanding based on confusing emergent non-reductionism with the impossibility to *understand*, represent, identify, and model processes and properties which are able to establish processes of emergence. We introduced possible ways by which processes of emergence take place within the conceptual framework of theoretical incompleteness, e.g. undecidability, and uncertainty given by equivalences, where everything must happen some way, and in some cases, in multiple ways.

Section 2 discussed the possible roles of compatibilities, equivalences, and their interplay and combinations at different levels within a framework of robustness, i.e. the ability to maintain and resume coherence

Section 3 discussed the role of emergence in complexity

Section 4 presented real processes considered as ways to establish processes of emergence, such as synchronisation, correlation, coherence, and polarisation

Section 5 discussed both processes compatible with and allow emergence and properties featuring processes of emergence and finally the constructivist role of the observer

The above should be understood as the study of signs and clues, of *theory-substitutive* concordances and correspondences, indicating and representing processes of emergence. *More realistically we should consider future contexts mixing theoretically symbolic and non-symbolic approaches allowing for soft-theorisations and incomplete-theorisations having, for instance, multiple dynamical incompleteness with possible partial coherences.*

Issues in Section 5 should not be interpreted through a reductionist paradigm simply as *component* parts, but as possibly dynamically dominant aspects which can take place in any combination and at any level.

Section 6 discussed issues related to simulations.

Finally, we stress the theoretical content and philosophical meaning of this approach finalised to represent coherent incompleteness of emergence and its non-symbolic representation as partially previously considered by connectionist theories based on artificial neural networks (da Silva et al. 2017).

We attempted to represent the dynamics of emergence in order to allow a better conceptual understanding of constituting processes which are not reducible to deterministic phenomena.

This is intended as a cultural contribution to many disciplinary contexts far from specialist approaches as mentioned in the introduction, such as architecture, economics, education, medicine and welfare, philosophy, politics, safety at work, and other general social issues.

Web Resources (Accessed on December 4, 2017)

Antipredator adaptation. https://www.revolvy.com/main/index.php?s=Antipredator%20 adaptation&item_type=topic.

Domowitz, I., & El-Gamal, M. (1997). *Financial market structure and the ergodicity of prices*. http://www.ssc.wisc.edu/econ/archive/wp9719.pdf.

Domowitz, I., & El-Gamal, M. (1999). *A consistent nonparametric test of ergodicity for time series with applications*. http://papers.ssrn.com/sol3/papers.cfm?abstract_id=179912.

Flock simulator. Retrieved from https://sourceforge.net/projects/msp3dfbsimulator/?source=directory. Complete information in: Minati, G. (2017). *Tentative guidelines for the implementation of meta-structural and network software models of collective behaviours* (pp. 16–22).Retrieved from http://arxiv.org/abs/1603.07174.

http://www.red3d.com/cwr/boids/.

Landon-Lane, J. S., & Quinn, J. A. (2000). *Growth and ergodicity: Has the world converged? The Econometric Society*. http://www.econometricsociety.org/meetings/wc00/pdf/0146.pdf.

McCaa, R. (2001) *An essay on inverse projection*. http://www.hist.umn.edu/~rmccaa/populate/ipessay.htm.

References

Acebrón, J. A., Bonilla, L. L., Vicente, C. J. P., Ritort, F., & Spigler, R. (2005). The Kuramoto model: A simple paradigm for synchronization phenomena. *Reviews of Modern Physics, 77*, 137–185.

Aihara, K., Imura, J., & Ueta, T. (Eds.). (2015). *Analysis and control of complex dynamical systems: Robust bifurcation, dynamic attractors, and network complexity*. New York: Springer.

Alligood, K. T., & Yorke, J. A. (2009). *Chaos: An introduction to dynamical systems*. New York: Springer.

Altamura, M., Elvevåg, B., Campi, G., De Salvia, M., Marasco, D., Ricci, A., et al. (2012). Toward scale-free like behavior under increasing cognitive load. *Complexity, 18*, 13–43.

Anderson, C. (2008). The end of theory: The data deluge makes the scientific method obsolete. *Wired magazine*, *16*(7). Retrieved December 4, 2017, from https://www.wired.com/2008/06/pb-theory/.

Anderson, P. W., & Stein, D. L. (1985). Broken symmetry, emergent properties, dissipative structures, life. Are they related? In F. E. Yates (Ed.), *Self organizing systems: The emergence of order* (pp. 445–457). New York: Plenum Press.

Baas, N. A. (1994). Emergence, hierarchies, and hyperstructures. In C. G. Langton (Ed.), *Alife III, Santa Fe Studies in the Sciences of Complexity, Proceedings* (Vol. XVII, pp. 515–537). Redwood City: Addison-Wesley.

Bailly, F., & Longo, G. (2011). *Mathematics and the natural sciences. The physical singularity of life*. London: Imperial College Press.

Baker, A. (2013). Complexity, networks, and non-uniqueness. *Foundations of Science., 18*, 687–705.

Ballerini, M., Cabibbo, N., Candelier, R., Cavagna, A., Cisbani, E., et al. (2007). Interaction ruling animal collective behavior depends on topological rather than metric distance: Evidence from a field study. *Proceedings of the National Academy of Sciences of the United States of America, 105*(4), 1232–1237.

Barabási, A. L. (2002). *Linked: The new science of networks*. Cambridge: Perseus Publishing.

Barbi, E., Bertino, S., & Sonnino, E. (Eds.). (2004). *Inverse projection techniques. Old and new approaches*. New York: Springer.

Batterman, R. (2011). Emergence, singularities, and symmetry breaking. *Foundations of Physics, 41*, 1031–1050.

Bohr, N. (1928). The quantum postulate and the recent development of atomic theory. *Nature, 121*(3050), 580–590.

Boccaletti, S. (2008). *The synchronized dynamics of complex systems*. Oxford: Elsevier.

Boccaletti, S., Kurths, J., Osipov, G., Valladares, D. L., & Zhouc, C. S. (2002). The synchronization of chaotic systems. *Physics Reports, 366*, 1–101.

Broersen, P. M. T. (2010). *Automatic autocorrelation and spectral analysis*. London: Springer.

Brown, J. H., & West, G. B. (2000). *Scaling in biology*. Oxford: Oxford University Press.

Bunde, A., & Havlin, S. (2012). *Fractals and disordered systems*. New York: Springer.

Buck, J., & Buck, E. (1966). Biology of synchronous flashing of fireflies. *Nature, 211*, 562–564.

Calude, C. C., & Longo, G. (2016). The deluge of spurious correlations in big data. *Foundations of Science, 22*(3), 595–612.

Cartwright, E. (2014). *Behavioral economics*. New York: Routledge.

Cavagna, A., Cimarelli, A., Giardina, I., Parisi, G., Santagati, R., et al. (2010). Scale-free correlations in starling flocks. *Proceedings of the National Academy of Sciences of the United States of America, 107*, 11865–11870.

Christensen, K., & Moloney, N. R. (2005). *Complexity and Criticality*. London, UK: Imperial College Press.

Cicogna, G. (1981). Symmetry breakdown from bifurcation. *Lettere al Nuovo Cimento, 31*, 600–602.

Cloude, S. (2014). *Polarisation: Applications in remote sensing*. Oxford: Oxford University Press.

Cohen, R., & Havlin, S. (2010). *Complex networks: Structure, robustness and function*. Cambridge: Cambridge University Press.

Coudène, Y. (2016). *Ergodic theory and dynamical systems*. London: Springer.

Cruchtfield, J. P. (1994). The calculi of emergence: Computation, dynamics and induction. *Physica D, 75*, 11–54.

da Silva, I. N., Hernane Spatti, D., Andrade Flauzino, R., Liboni, L. H. B., & dos Reis Alves, S. F. (2017). *Artificial neural networks: A practical course*. Basel: Springer.

De Wolf, T., & Holvoet, T. (2005). Emergence versus self organisation: Different concepts but promising when combined. In S. A. Brueckner, G. Di Marzo Serugendo, A. Karageorgos, & R. Nagpal (Eds.), *Engineering self-organising systems: Methodologies and applications, Lecture notes in computer science* (Vol. 3464, pp. 1–15). Berlin: Springer.

De Wolf, T., Holvoet, T., & Samaey, G. (2006). Development of self-organising emergent applications with simulation-based numerical analysis. In S. A. Brueckner, G. Di Marzo Serugendo, & A. Karageorgos (Eds.), *Engineering self-organising systems* (pp. 138–152). Berlin: Springer.

De Wolf, T., Samaey, G., & Holvoet, T. (2005). Engineering self-organising emergent systems with simulation-based scientific analysis. In S. Brueckner, G. Di Marzo Serugendo, D. Hales, & F. Zambonelli (Eds.), *Proceedings of the Third International Workshop on Engineering Self-Organising Applications* (pp. 146–160). New York: Springer.

Doncaster, C. P., & Davey, A. J. H. (2007). *Analysis of variance and covariance*. Cambridge: Cambridge University Press.

Dorogovtsev, S. N., Goltsev, A. V., & Mendes, J. F. F. (2008). Critical phenomena in complex networks. *Reviews of Modern Physics, 80*, 1275–1335.

Duncan, P., & Vigne, N. (1979). The effect of group size in horses on the rate of attacks by blood-sucking flies. *Animal Behaviour, 27*, 623–625.

Estrada, E. (2016). *The structure of complex networks: Theory and applications*. Oxford: Oxford University Press.
Everitt, B. S., & Landau, S. (2011). *Cluster Analysis*. Chichester: Wiley.
Falconer, K. (2013). *Fractals: A very short introduction*. New York: Oxford University Press.
Fernandez, N., Maldonado, C., & Gershenson, C. (2014). Information measures of complexity, emergence, self-organization, homeostasis, and autopoiesis. In M. Prokopenko (Ed.), *Guided self-organization: Inception. Emergence, complexity and computation* (Vol. 9, pp. 19–51). Berlin, Heidelberg: Springer.
Fields, C. (2016). Building the observer into the system: Toward a realistic description of human interaction with the world. *Systems, 4*(4), 32 Retrieved December 4, 2017, from http://www.mdpi.com/2079-8954/4/4/32.
Foster, W. A., & Treherne, J. E. (1981). Evidence for the dilution effect in the selfish herd from fish predation on a marine insect. *Nature, 295*, 466–467.
Freeman, W. J., Kozma, R., & Werbos, P. (2001). Adaptive behavior in complex stochastic dynamical systems. *BioSystems, 59*, 109–123.
Gabbay, D. M., & Woods, J. (2005). *A practical logic of cognitive systems: The reach of abduction: Insight and trial*. Amsterdam: Elsevier.
Gambuzza, L. V., Cardillo, A., Fiasconaro, A., Fortuna, L., Gómez-Gardeñes, J., & Frasca, M. (2013). Analysis of remote synchronization in complex networks. *Chaos, 23*(4), 1–8.
Getling, A. V. (1998). *Rayleigh-Bénard convection: Structures and dynamics*. Singapore: World Scientific.
Gilboa, I., Samuelson, L., & Schmeidler, D. (2015). *Analogies and theories: Formal models of reasoning*. Oxford: Oxford University Press.
Giuliani, A. (2014). Networks as a privileged way to develop mesoscopic level approaches in systems biology. *Systems, 2*(2), 237–224 Retrieved December 4, 2017, from http://www.mdpi.com/2079-8954/2/2/237.
Giuliani, A., & Zbilut, J. (2008). *The latent order of complexity*. New York: NovaScience.
Goosse, H. (2015). *Climate system dynamics and modelling*. New York: Cambridge University Press.
Haken, H. (1983). *Advanced synergetics*. Berlin: Springer.
Hair Jr., J. F., & Black, W. C. (2013). *Multivariate data analysis*. Harlow: Pearson.
Heard, D. (2006). A new problem for ontological emergence. *Philosophical Quarterly, 56*, 55–62.
Heisenberg, W. (1971). *Physics and beyond*. New York: Harper & Row.
Hemelrijk, C. K., & Hildenbrandt, H. J. (2015). Scale-free correlations, influential neighbours and speed control in flocks of birds. *Journal of Statistical Physics, 158*, 563–578.
Henriksen, R. N. (2015). *Scale invariance: Self-similarity of the physical*. Weinheim: Wiley.
Hirst, L. S. (2013). *Fundamentals of soft matter science*. Boca Raton: CRC Press.
Hong, H., Park, H., & Choi, M. Y. (2005). Collective synchronization in spatially extended systems of coupled oscillators with random frequencies. *Physical Review E, 72*, 036217–036218.
Jakus, G., Milutinovic, V., Omerovic, S., & Tomazic, S. (2013). *Concepts, ontologies, and knowledge representation*. Springer Briefs in Computer Science. New York: Springer.
Jolliffe, I. T. (2002). *Principal component analysis*. New York: Springer.
Kinoshita, S. (Ed.). (2013). *Pattern formations and oscillatory phenomena & Belousov-Zhabotinsky reaction*. Amsterdam: Elsevier.
Krause, J., & Ruxton, G. D. (2002). *Living in groups*. New York: Oxford University Press.
Kumar, K. (2004). *From Post-Industrial to Post-Modern Society: new theories of the contemporary world*. Oxford: Blackwell Publishers.
Kuznetsov, Y. (2004). *Elements of applied bifurcation theory*. New York: Springer.
Laughlin, R. B., Pines, D., Schmalian, J., Stojkovic, B. P., & Wolynes, P. (2000). The middle way. *Proceeding of the National Academy of Sciences of the United States of America, 97*(1), 32–37.
Lee, R. D. (1974). Estimating series of vital rates and age structures from baptisms and burials: A new technique with applications to pre-industrial England. *Population Studies, 28*, 495–512.
Lee, R. D. (1978). *Econometric studies of topics in demographic history*. New York: Arno Press.

Lee, R. D. (1985). Inverse projection and back projection: A critical appraisal and comparative results for England, 1539-1871. *Population Studies, 39*, 233–248.

Lewis, T. G. (2009). *Network science: Theory and applications*. Hoboken: Wiley.

Licata, I. (2008). Logical openness in cognitive models. *Epistemologia, 31*, 177–192.

Licata, I. (2012). Seeing by models: Vision as adaptive epistemology. In G. Minati, M. Abram, & E. Pessa (Eds.), *Methods, models, simulations and approaches towards a general theory of change* (pp. 385–400). Singapore: World Scientific.

Licata, I., & Minati, G. (2016). Emergence, computation and the freedom degree loss information principle in complex systems. *Foundations of Science, 21*, 1–19.

Liljenstrom, H., & Svedin, U. (Eds.). (2005). *Micro, meso, macro: Addressing complex systems couplings*. Singapore: World Scientific.

Lind, P. G., Gallas, J. A. C., & Herrmann, H. J. (2009). Coherence in complex networks of oscillators. In M. Ausloos & M. Dirickx (Eds.), *The logistic map and the route to chaos*. Berlin: Springer.

Longo, G., & Montévil, M. (2014). Perspectives on Organisms. Biological time, Symmetries and Singularities. New York, NY: Springer.

Lorenz, E. (1963). Deterministic non period flow. *Journal of the Atmospheric Sciences, 20*, 130–141.

Magurran, A. E. (1990). The adaptive significance of schooling as an anti-predator defence in fish. *Annales Zoologici Fennici, 27*(2), 51–66.

Manrubia, S. C., & Mikhailov, A. S. (2004). *Emergence of dynamical order: Synchronization phenomena in complex systems*. Singapore: World Scientific.

Minati, G. (2012a). Knowledge to manage the knowledge society. *The Learning Organisation, 19*(4), 352–370.

Minati, G., (2012b). (Guest editor). Special issue: Knowledge to manage the knowledge society. *The Learning Organisation, 19* (4), 296.

Minati, G., & Licata, I. (2015). Meta-structures as multidynamics systems approach. some introductory outlines. *Journal on Systemics, Cybernetics and Informatics, 13*(4), 35–38 Retrieved December 4, 2017, from http://www.iiisci.org/journal/sci/issue.asp?is=ISS1504.

Minati, G., Licata, I, & Pessa, E. (2013). Meta-structures: The search of coherence in collective behaviours (without physics); In A. Graudenzi, G. Caravagna, G. Mauri, & M. Antoniotti (Eds.), *Proceedings Wivace 2013*, Milan, Italy, July 1–2 (Vol. 130, pp. 35–42). Retrieved December 4, 2017, from http://rvg.web.cse.unsw.edu.au/eptcs/paper.cgi?Wivace2013.6.

Minati, G., & Licata, I. (2013). Emergence as mesoscopic coherence. *Systems, 1*(4), 50–65.

Minati, G., & Licata, I. (2012). Meta-structural properties in collective behaviours. *International Journal of General Systems, 41*, 289–311.

Minati, L. (2015). Remote synchronization of amplitudes across an experimental ring of non-linear oscillators. *Chaos, 25*(12), 123107–123112.

Minati, L., Chiesa, P., Tabarelli, D., D'Incerti, L., & Jovicich, J. (2015). Synchronization, non-linear dynamics and low-frequency fluctuations: Analogy between spontaneous brain activity and networked single-transistor chaotic oscillators. *Chaos, 25*(3), 033107–033127.

Minati, L., de Candia, A., & Scarpetta, S. (2016). Critical phenomena at a first-order phase transition in a lattice of glow lamps: Experimental findings and analogy to neural activity. *Chaos, 26*(7), 073103–073111.

Minati, G. (2016a). Knowledge to manage the knowledge society: The concept of *theoretical incompleteness*, Systems, 4(3), 26. Retrieved December 4, 2017, from http://www.mdpi.com/2079-8954/4/3/26.

Minati, G. (2016b). General System(s) Theory 2.0: A brief outline. In G. Minati, M. Abram, & E. Pessa (Eds.), *Towards a post-bertalanffy systemics, Proceedings of the Sixth National Conference of the Italian Systems Society, Rome, Italy, 21–22 November 2014* (pp. 211–219). New York: Springer.

Minati, L., Winkel, J., Bifone, A., Oświęcimka, P., & Jovicich, J. (2017). Self-similarity and quasi-idempotence in neural networks and related dynamical systems. *Chaos, 27*(4), 043115.

Minati, G., & Pessa, E. (2006). *Collective beings*. New York: Springer.
Minati, G., & Pessa, E. (2018). *From collective beings to quasi-systems*. New York: Springer.
Missaoui, R., & Sarr, I. (2015). *Social network analysis - Community detection and evolution*. New York: Springer.
Mitra, P. (2014). *Symmetries and symmetry breaking in field theory*. Boca Raton: CRC Press.
Motter, A. E., & Albert, R. (2012). Networks in motion. *Physics Today, 65*, 43–48.
Newman, M. (2010). *Networks: An introduction*. New York: Oxford University Press.
Newman, M., Barabasi, A.-L., & Watts, D. J. (Eds.). (2006). *The structure and dynamics of networks*. Princeton: Princeton University Press.
Nitzan, A., & Ortoleva, R. (1980). Scaling and Ginzburg criteria for critical bifurcations in nonequilibrium reacting systems. *Physical Review A, 21*, 1735–1755.
Paine, A. D. M. (1985). Ergodic reasoning in geomorphology: time for a review of the term? *Progress in Physical Geography, 9*, 1–15.
Paperin, G., Green, D. G., & Sadedin, S. (2011). Dual-phase evolution in complex adaptive systems. *Interface, 8*, 609–629.
Pessa, E. (2006). Physical and biological emergence: Are they different? In G. Minati, E. Pessa, & M. Abram (Eds.), *Systemics of emergence. Research and development* (pp. 355–374). Berlin: Springer.
Pessa, E. (2012). On models of emergent metastructures. In G. Minati, E. Pessa, & M. Abram (Eds.), *Methods, Models, simulations and approaches towards a general theory of change* (pp. 113–132). Singapore: World Scientific.
Pessa, E. (1998). Emergence, self-organization, and quantum theory. In G. Minati (Ed.), *Proceedings of the First Italian Conference on Systemics* (pp. 59–79). Milan: Apogeo scientific.
Pourahmadi, M. (2013). *High-dimensional covariance estimation*. Hoboken: Wiley.
Prigogine, I. (1967). Dissipative structures in chemical systems. In S. Claesson (Ed.), *Fast reactions and primary processes in chemical kinetics* (pp. 371–382). New York NY: Interscience.
Prigogine, I. (1981). *From being to becoming: Time and complexity in the physical sciences*. New York: W. H. Freeman.
Prigogine, I. (1998). *The end of certainty: Time, chaos, and the new laws of nature*. New York: The Free Press.
Pulliam, H. R. (1973). On the advantages of flocking. *Journal of Theoretical Biology, 38*, 419–422.
Reynolds, C. (1987). Flocks, herds, and schools: A distributed behavioral model. *Computer Graphics, 21*, 25–34.
Roberts, M., & Russo, R. (1999). *A student's guide to analysis of variance*. New York: Routledge.
Rosen, M. (1982). *Hegel's dialectic and its criticism*. New York: Cambridge University press.
Rossetti, M. D. (2015). *Simulation modeling and arena*. Hoboken: Wiley-Blackwell.
Ruelle, D. (1989). *Chaotic evolution and attractors*. New York: Cambridge University Press.
Samaey, G., Holvoet, T., & De Wolf, T. (2008). Using equation-free macroscopic analysis for studying self-organising emergent solutions. In *Proceedings of the Second IEEE International Conference on Self-Adaptation and Self-Organization, IEEE Computer Society* (pp. 425–434), Venice, Italy, October 20–24.
Schroeder, M. (2009). *Fractals, chaos, power laws: Minutes from an infinite paradise*. New York: Dover Publications.
Shevlyakov, G. L., & Oja, H. (2016). Robust Correlation: Theory and Applications. Chichester, UK: Wiley.
Sparrow, C. (2013). *The Lorenz equations: Bifurcations, chaos, and strange attractors*. New York: Springer.
Stanley, H. E., Amaral, L. A. N., Gopikrishnan, P., Ivanov, P. C., Keitt, T. H., & Plerou, V. (2000). Scale invariance and universality: Organizing principles in complex systems. *Physica A: Statistical Mechanics and its Applications, 281*, 60–68.
Steffe, L. P., & Thompson, P. W. (Eds.). (2010). *Radical constructivism in action: Building on the pioneering work of Ernst von Glasersfeld*. London: Routledge.

Stein, D. L. (1980). Dissipative structures, broken symmetry, and the theory of equilibrium phase transitions. *Journal of Chemical Physics, 72*, 2869–2874.
Strocchi, F. (2010). *Symmetry breaking*. Berlin: Springer.
Taylor, S. (Ed.). (2014). *Agent-based modeling and simulation*. New York: Palgrave Macmillan.
Transtrum, M. K., Machta, B. B., Brown, K. S., Daniels, B. C., Myers, C. R., & Sethna, J. P. (2015). Perspective: Sloppiness and emergent theories in physics, biology, and beyond. *The Journal of Chemical Physics, 143*(1), 010901–010913.
Turing, A. (1936). On computable numbers, with an application to the Entscheidungsproblem. *Proceedings of the London Mathematical Society, 42*, 230–265.
Urdan, T. C. (2016). *Statistics in plain English*. New York: Routledge.
Uzunov, D. I. (2010). *Introduction to the theory of critical phenomena: Mean field, fluctuations and renormalization*. Singapore: World Scientific.
Valente, T. W. (2012). Network interventions. *Science, 337*, 49–53.
Vicsek, T., & Zafeiris, A. (2012). Collective motion. *Physics Reports, 517*, 71–140.
Von Bertalanffy, L. (1968). *General system theory: Foundations, development, applications*. New York: George Braziller.
Webber Jr., C. L., Ioana, C., & Marwan, N. (Eds.). (2016). *Recurrence plots and their quantifications: Expanding horizons*. New York: Springer.
Webber Jr., C. L., & Marwan, N. (Eds.). (2016). *Recurrence quantification analysis*. New York: Springer.
Wilson, E. O. (1975). *Sociobiology*. Cambridge: Belknap/Harvard University Press.
Wu, J. (2012). *Advances in K-means clustering: A data mining thinking*. Berlin: Springer.

The World Opacity and Knowledge

Giuseppe Vitiello

Abstract In a typical scattering process among elementary particles, observation is limited to the asymptotic regions, where ingoing particles and outgoing particles behave like free, non-interacting particles. The region of interaction is not accessible to our observations. These would be interfering with the phenomenon under study. The interaction region is thus an "opacity" region for us. Starting from such a remark, I then discuss the behavior of open systems and their interaction with the environment. The discussion is further extended to the brain functional activity and to the possibility to describe consciousness and mental activity as inseparably linked to neuronal activity.

1 Introduction

The study of elementary particle physics and condensed matter exhibits some formal and methodological features susceptible to be recognized also in the study of other disciplines. On the one hand, methodological tools in the study of physics can be extended to different disciplines due to their general validity, on the other, the same dynamical approach happens to be useful in order to analyze the physical basis underlying, for example, the richness of the biochemical and cellular phenomenology in biological systems. In this chapter I consider the notion of *opacity* in the observation of scattering processes in elementary particle physics and I extend the discussion so to include systems in interaction with the environment in which they are embedded, as typically it happens to living systems. I consider in particular the brain functional activity from the perspective of the dissipative quantum model of brain and report about some results of studies of consciousness and mental phenomena. Apart specific motivations coming from the general relevance of these subjects,

G. Vitiello (✉)
Dipartimento di Fisica "E.R.Caianiello", Università di Salerno and Istituto Nazionale di Fisica Nucleare, Salerno, Italy
e-mail: vitiello@sa.infn.it

L. Urbani Ulivi (ed.), *The Systemic Turn in Human and Natural Sciences*, Contemporary Systems Thinking, https://doi.org/10.1007/978-3-030-00725-6_2

I had already the occasion to motivate, in a past publication (Vitiello 2004), the need to pursue a unitary conception of knowledge, which has been always present in our cultural inheritance, surviving also at present time, although struggling against the blind ideology pursuing "profit above all." One of the most lucid expression of such an effort in reaching a unified view of nature has been provided by Lucretius in his *De rerum natura* (Titus Lucretius Carus (99–55 B.C.) 2003):

> "...we must not only give a correct account of celestial
> matter, explaining in what way the wandering of the sun
> and moon occur and by what power things happen on earth.
> We must also take special care and employ keen reasoning
> to see where the soul and the nature of the mind come from,..."

The plan of the chapter is the following. In Sects. 2 and 3, I introduce the notion of opacity in metals and other bodies interacting with light. In Sects. 4–6, I consider general features of elementary particle scattering processes, the relation between the level of the observables and the level of dynamical processes and I comment on perturbative and non-perturbative physics, on open and closed systems. Sections 7–10 are devoted to the inclusion in the discussion of biological system and the brain, closing with remarks on the role of time, consciousness, and the mind activity. I want to express my gratefulness to F. Desideri and P.F. Pieri for allowing me to report in this chapter the translation of the paper on a similar subject published in the magazine Atque (Vitiello 2016a).

2 Opacity and Transparency

Opacity is an optical property characterizing the behavior of a metal interacting with light (Amaldi 1962; Rossi 1957). It consists of the partial or total absorption of a beam of light (electromagnetic[1] (em) wave) hitting a body. If there is no absorption, the body is said to be transparent; light passes through it without losing energy, as the body would not be on its path, *invisible* to it. In fact, something *visible*, opaque, indeed, is sometimes put on the glass of a window or a door, otherwise dangerously transparent.

Opacity and transparency can be considered as a response of the body to the interaction with light, deriving, on the one hand, from the behavior of the body elementary components and structural properties, on the other from the intensity and frequency of light. One can have different responses to different light intensities and frequencies, also depending on whether the body has, for example, a crystalline or amorphous structure. Opacity and transparency are therefore not intrinsic properties of bodies. They describe the way the body "manifests" when using light as an instrument of observation.

[1] Light is an em radiation; in the following the word light is used to denote generically an em wave and vice-versa.

3 Intrinsic Opacity

Things are actually a bit more complex. What happens is that when the light and the body "interact" there is a whole series of phenomena not directly observable, but which are crucial for the final manifestation of the opacity or transparency of the body in question.

Although, as we have seen, opacity and transparency are associated with being the body visible or, respectively, invisible to a beam of light, there is an *intrinsically* invisible dynamic level, impenetrable to our observations; namely the level of the dynamical microscopic processes, which would be strongly disturbed, or even destroyed, by our observations.[2]

It is perhaps this level inaccessible to our direct observation that we can refer to as a *substantial or intrinsic opacity* of microscopic phenomena.

An em wave is characterized by a specific *frequency*, that is, by the number of oscillations per second of the electric field associated to the wave, and by the *wavelength*. Frequency and wavelength are linked to each other by the propagation speed of the em wave which in the vacuum is the speed of light. The wave can be absorbed to a greater or lesser extent when it propagates in a body (Amaldi 1962; Rossi 1957).

Consider the case of a metal. The electric field that propagates with the wave interacts with the electrons present in the metal generating an electric current that subtracts energy from the wave dissipating it in the form of heat.[3] The interaction of the wave with the electrons of the metal is however possible if and only if the electrons can oscillate with the same frequency of the wave. It is necessary that wave and electrons enter into resonance. Only in this case the wave can drag the electrons with it in its propagation, generating the electric current. The energy that the electric current takes away from the wave depends on various factors, for example the thickness of the metal, its conductivity, temperature, etc. Hence the greater or less wave energy attenuation during the passage of the wave (of light) through the metal, the greater or less opacity of the metal. For waves of very high frequencies it happens that the electrons, due to the inertia coming from their mass, small but not zero, cannot follow the rapid variations of the electric field. In such case, electric current is not generated and there is no dissipation of energy of the wave, which passes undisturbed through the metal. *The wave does not see the metal* (and the metal does not see the wave).

In the case of insulating bodies, the atoms and the molecules, that compose them, are the ones which can oscillate at frequencies equal or similar to those of the waves that invest them, with consequent absorption or transmission phenomena. Typically,

[2] We do not enter here in the discussion of the role of the observer in determining the result of the observation of microscopic and quantum phenomena, which goes beyond the scope of this work.

[3] This dissipation process goes under the name of *Joule effect*. We are sure that the reader knows what the Joule effect is. In fact, he knows that the current that circulates in the resistance of an electric stove or in the filament of an incandescent lamp generates heat. The lamp, like the stove, heats up and it is good not to touch it ... This is the Joule effect.

the absorptions also determine the color of the various substances. It is determined by the frequencies not absorbed, or reflected (also called complementary to the absorbed frequencies), when these are in the spectrum of visible light.

4 Observables and Dynamics

In describing the interaction of an em wave with a body, it has been mentioned that there is an *opaque* region, inaccessible to our observation, which otherwise would produce a destructive interference with the phenomenon we want to observe. In order to identify general aspects of an observation process, it is useful to consider the description of the interaction between elementary components of a system, as schematized for example in the physics of elementary particles[4] and represented in Fig. 1.

There, three regions can be distinguished. Starting from the left, in the "input" region the "incoming particles" can be identified with appropriate detectors. These must be placed at a distance from the central region of the "dynamics," in such a way to avoid any possible interference with the interaction whose effects are to be studied, and also such that the incoming particles can be considered "free," "not yet" interacting with each other, "isolated" from each other. This request is justified by the fact that if we want to study the interaction between the particle A and the particle B, we must be certain that the particles entering the interaction region are actually A and B. The identification of the incoming particles must therefore be made far in space from the dynamic region and even long before the time when their interaction takes place. Finally, in the "output" region the detectors are arranged to identify what the interaction generated, the "outgoing particles." Also these detectors must be placed "far" in space and time from the interaction region, always to

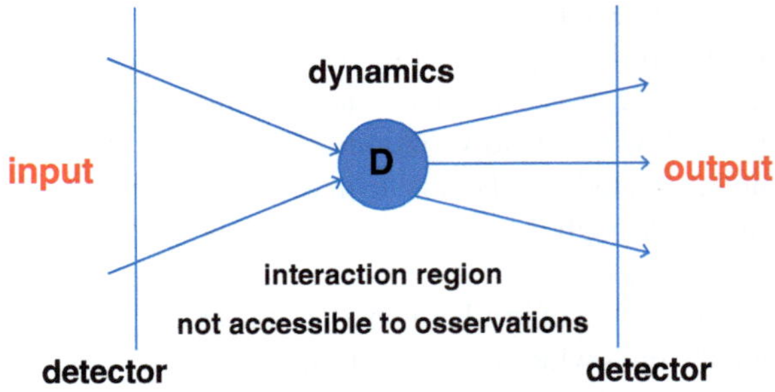

Fig. 1 Schematic representation of particle interaction

[4]In addition to the em interaction we know the gravitational interaction, the weak one (about 1000 times weaker than the em interaction), responsible for, for example, the process of decay of the neutron and the strong one (about 1000 times stronger than the em interaction) responsible for the interaction between proton and neutron, particles of which the atomic nucleus is composed (in reality the reference force is that between the quarks which are the constituents of protons and neutrons).

avoid interference with the dynamic region and to be able to identify the products of the interaction, which must therefore be "free," "no longer" interacting with each other. From now on, the "in" and "out" regions, due to their distance in space and time from the interaction region, will be called "asymptotic regions."

It is easy to be convinced that this schematization is actually generalizable to many other situations, even different from the case of interactions between particles. It shows us that we have actually two "levels" of description, that of the observable particles in the asymptotic regions and the "dynamic" level, inaccessible to our observation. However, the schematization rests on some assumptions that need to be clarified and discussed, which we will do later. Here we see that, all in all, the interesting part of the game, the dynamic region, where there are "events," remains *invisible* to us, intrinsically *opaque*. Access is granted to us only in the asymptotic regions, where "things" are detected by us (perceived) as non-interacting. On what happens in the dynamic region we can only formulate *theoretical hypotheses*. The *credibility* of these hypotheses (theory) is entrusted to their *verifiability* in the comparison between theoretical *predictions*, based on observations in the "in" region and on calculations according to the theory, and observations in the "out" region.

5 The Ontological Prejudice

The schematization introduced above plays a fundamental role in physics and in general in all scientific studies and their applications. It is also very useful in our daily behavior, in our judgements, forecasts, and decisions.

The mathematical formalism we have, called canonical formalism, is based on it. The assumption, or fundamental postulate on which it rests consists in the fact that it is actually possible to "isolate," for the purpose of their identification, each of the incoming and outgoing particles (here and below we will continue to call "particle" the entity or the system object of our interest). As already noted, this requires that they are "non-interacting" with each other. In other words, it is assumed that it is possible to "switch off" their interaction in asymptotic regions and "turn it on" in the dynamic region. In reality, this is a double postulate, since it is also assumed that particles can have their existence, independent and in the absence of any possible interaction; they can be "existing by themselves."

As we will see later, this postulate is however only a *prejudice*, which we refer to as the *ontological prejudice*. Obviously *we need*, for a matter of psychic stability, to believe that our very identity is independent of our interaction with the world and with others. We *want* to believe that we remain ourselves even when the world and the others do not exist. We therefore extend our *belief* to what surrounds us, to others, to all things, to the world, which therefore consists of a set of "identities" disjoined among themselves, "free" from any bond of mutual interaction; it is then assumed that one is able to "choose" and "adjust" the interactions with the world; in practice, we find ourselves in a constant struggle to cut unpleasant, unwanted constraints, to ward off presences we believe we can do without.

In the following the limits of real applicability of this double postulate will be discussed. Here we observe that the possibility, in many cases effective and efficacious, of adjusting or controlling the forces around us let us to consider the onset of a force, previously "switched off" in the asymptotic "in" region, as a *perturbation* to the non-interacting state. It is then possible to think of the transition from the non-interacting state to the one in which the interaction is "completely switched on" (in the dynamic region, Fig. 1) as a succession of increasingly strong perturbations to which the free state of the particle undergoes in its evolution. The mathematical apparatus that describes such a process takes the name of perturbation theory and the corresponding physics is called perturbative physics. The computational power and the predictive capacity that offers such an approach has proved to be enormous. It is easy to convince ourselves that even in our daily activities we proceed according to this *perturbative scheme*, attributing to weak forces (small perturbations) weak effects (or even negligible effects to negligible forces).

6 The Naive Vision of the World

In physics, the perturbative scheme discussed so far is not the only possible one. There is a whole set of systems, whose study gave rise to the physics of "nonlinear dynamical systems," in which it is observed that not necessarily weak forces generate weak effects. There are in fact numerous examples, in physics as in biology, in chemistry, etc., in which it happens the opposite to what happens in the perturbative scheme, namely that remarkable effects or even catastrophic ones are caused by weak interactions, or as one uses to say, by weak couplings. Nonlinearities in mathematics occur for self-interacting systems or when they interact with other systems giving rise to anomalous enhancements in the interaction amplitudes. In simple words, in linear dynamics enhancements in the interaction amplitudes are due to "overlaps" or sums, in nonlinear dynamics they are due to "products." From a technical point of view, nonlinear solutions depend on the intensity of the interaction λ, called coupling constant, through a positive power of $1/\lambda$ and therefore λ can never be equal to zero. This means that it is never possible to eliminate, to "switch off" the interaction (which would happen for $\lambda = 0$), i.e., it is never possible to have free objects (particles), subtracted from any interaction. In such cases, the asymptotic regions "in" and "out" in Fig. 1 simply do not exist. Only the dynamic region exists. The perturbative approach loses its meaning and leaves room for the *non-perturbative physics*. The vision of a world made up of parts, entities isolated from each other, existing by themselves, which may or may not enter into mutual interaction, is not conceivable in a world of nonlinear phenomena. This appears to be *the naive vision of the world* ("der naiven Weltansicht"; Cassirer 1920). The limits of validity of the perturbative approach become evident also due to the discovery of quantum phenomena where a crucial role is played by the quantum fluctuations of the system's minimum energy state, called vacuum. Such fluctuations and their interactions with quantum systems can never be eliminated. The example of quarks, the elementary components of the sub-nuclear particles is paradigmatic in this sense. Quarks are permanently *confined* within the particles of which they are the components. There are no free, asymptotic quarks; it is not possible to turn off the interaction that binds them. In this sense, it can be said that quarks are not existing in the absence of their interaction. They are not the "subjects," the source of the interaction. Indeed, it is the interaction that "defines" the interacting objects, not the other way around.[5]

Given the *fundamental* nature of the quantum world, it turns out that the non-perturbative vision is the one valid at the fundamental level and the perturbative approach appears to be only a *convenient* one in order to make the calculations and the mathematical apparatus simple.

The vision that we then reach is the one of systems whose interaction with other systems can never be switched off, systems *open* on the world in which they are immersed.

7 Open Systems and Closed Systems

An open system is in continuous exchange of energy, matter, momentum, etc. with the world in which it is embedded. Open systems are therefore also called dissipative systems. The world, or environment, has the role of "reservoir" for the system. In it the system finds the source of energy to draw from, if it needs energy in its evolution, and the deposit in which to pour the energy it needs to dispose of.

On the other hand, the mathematical formalism available to us, called *canonical* formalism, is actually limited to *closed* systems, those that can be considered as isolated, subtracted from any interaction with other bodies and systems.

Let me denote generically with A the variables (or degrees of freedom) that describe our open system and with $\tilde{A}$ the degrees of freedom of the environment in which it is immersed. With the tools offered by the canonical formalism, it is then impossible to study the open system A, without considering at the same time its environment $\tilde{A}$. We need to consider both A and $\tilde{A}$, in such a way that *the whole* $\{A, \tilde{A}\}$ constitutes a closed system. The canonical formalism can be then applied to it. From the point of view of the balance of flows, for example of energy, between A and $\tilde{A}$, environment receives from A all that from A leaves, and yields to A all that A receives: "out" for A is "in" for $\tilde{A}$, and vice-versa. In formal terms, the exchange "in $\leftrightarrow$ out" is described by inverting the relative sign of time, for example by inverting it in the description of $\tilde{A}$. This then turns out to be the "inverted in time" image of A, its image in the "time mirror," its *Double* (Celeghini et al. 1992; Vitiello 1995, 2001). All this is formally described by the algebraic formalism which "doubles the degrees of freedom": $A \rightarrow \{A, \tilde{A}\}$. The state of the whole system $\{A, \tilde{A}\}$ turns out to be a coherent state in which A and $\tilde{A}$ are reciprocally coupled (entanglement). The predictive power of the theoretical apparatus is verified by the collective behavior of the system $\{A, \tilde{A}\}$.

[5]It is a bit like the theater and the literature: it is the plot of the show or of the novel that defines the characters. In the absence of plot, no character exists.

8 The Arrow of Time

Let me now consider the role of time. For closed systems, contrary to what happens for open systems, the energy of the system is a conserved quantity. There are no losses or energy gains for the system (which otherwise would not be closed). In the canonical formalism the conservation of energy is derivable, according to the Noether theorem, from the symmetry of the dynamics under transformations that translate the time variable by a constant c: $t \rightarrow \tau = t + c$. In other words, the equations describing the evolution in time of a closed system are symmetric under time translations; they do not change their formal aspect (their form) when the position of the origin on the time axis is shifted by a constant quantity. The operation is similar to what happens by adjusting the clocks 1 h forward or backward, when introducing legal or saving time or returning to the solar one. In the case of closed systems, the exact position of the origin on the time axis does not have therefore a substantial value. It can be moved at will. Time does not have an absolute value, only time intervals are important because they remain unchanged under translation of the origin of the time axis: $\tau - \tau' = (t + c) - (t' + c) = t + c - t' - c = t - t'$. In such a situation there is no notion of the present, of "now," "at this moment," nor the past or future are definitely distinct. There are therefore no clocks to synchronize, nor is there a unique "direction of time" since the origin of time can also be moved "backwards." There is no *history*, neither beginning, nor end. The flow of time destroys every origin that can be fictitiously assigned to its axis. It's like …Kronos had eaten his sons… (Vitiello 2016b, c).

Such a picture changes for open systems. For these, dissipation implies that energy is not conserved and there is no symmetry under spatial translation. As said, due to the Noether's theorem, symmetry under spatial translation implies energy conservation, and this does not happen for open systems. We have therefore that the origin on the temporal axis cannot be translated at will, is fixed. It marks the "birth" of our system and it cannot be changed. Memory becomes then possible, now it makes sense to recall. The dissipative system has a history, it ages and has a life time. It needs his Double for mutual energy exchanges. There are no arbitrary clocks. The Double acts like the watch for the system; it keeps track of the flow of time, whose direction, *the arrow of time*, is now not reversible. "It is the revenge of the sons of Kronos" (Vitiello 2016b, c).

9 The Brain Action-Perception Cycle

We started from simple physical considerations on the definition of opacity in metals and insulating materials and the existence of "regions of opacity" has been recognized to be intrinsic to observation processes. We have also realized that this also happens in the study of open systems. Our discussion will be now extended to some aspects of our relationship with the world around us.

The activity of the brain, which is indeed an open system, is characterized by *the action-perception cycle* (Freeman 1975; Merleau-Ponty 1942) in its interaction with the world. The brain places itself in the environment by formulating hypotheses and subjecting them to verification with intentional actions, constructing in this way knowledge through trial-and-error steps (Vitiello 2008). The stimuli received through the perceptive channels are framed in the landscape of previous perceptual experiences and in this process the net of correlations among them is enriched and renewed, "meanings" are thus constructed out of information. Each new perception is not simply added to the perceptual experience already acquired, as it happens, for example, for a new item added to a dictionary. In the case of the brain, each new perception changes the entire landscape of the meanings constructed up to then. Memory is not memory of information, it is memory of meanings.

The discussion on open systems in the previous sections extends to the brain which thus also appears inextricably linked to its Double, one open to the other. Their relationship is a dynamic one and "opaque," never accessible to an external observer, a *first person* experience. The act of consciousness lies in the dialogue between the brain (*the self*) and its Double (Vitiello 1995, 2001; Desideri 2011, 2018), all within their relationship, always new in its dynamic being. This might appear to constitute a problem from the stand point of scientific methodology, which requires that every phenomenon must be accessible to every observer, no matter where or when he carries his measurements. However, one should remember that such a methodology has been formulated for the study of closed systems and is limited to them. On the contrary, consciousness is framed in the context of open systems, for which, as discussed in previous sections, time plays a privileged role and their phenomenology is characterized by the existence of "opaque" dynamic regions.

10 Mind and Brain

The "hypotheses" formulated by the brain in the action-perception cycle are due to the Double. They constitute the "anticipatory" vision that the brain makes of the world, which, anticipating "from the future situations" that in the future may occur, determines the activity of the brain motor centers, with the consequent actions that the body undertakes. This is what commonly happens in the constant behavioral control; think, for example, of driving a vehicle that requires a "pre-vision" capability of what will happen on the road we travel.

It is therefore the Double that formulates the hypotheses. These do not belong to the baggage of memories, but to the imagination (vision) that projects backwards in time ("in the past" with respect to the expected event) what "is about to happen." This is "the mind" (Freeman and Vitiello 2016). On this basis the brain exerts its control over the actions to be taken.

In neuroscience studies, it is not an easy task to comprehend "mental activity" in terms of neuronal activity. In fact, there are different lines of thought (Atmanspacher

2015). There is a "dualist" vision that postulates a double level, the one of matter, physical or strictly neuronal, and the ideal level, separated from the physical level, to which mental activity belongs, in this sense a meta-physical activity. The strictly idealistic vision, in which the neuronal activity "derivates" from the "mind," is not independent of, but conditioned by it. The "unicist" vision, in which there is no possibility to distinguish between mental activity and neuronal activity. Then there is a fine spectrum of distinctions within each position. On the other hand, there is also a different hypothesis (Freeman and Vitiello 2016) rooted in the study of the brain as a physical system, in its full complexity of its cellular and biochemical components, characterized, however, by its being an open, dissipative system, whose study requires the introduction of the Double as a requirement dictated by physics and mathematics.

In the quantum dissipative model, the environment in which the brain is immersed forms the reservoir which provides the sink and the source of the energy exchanged by the brain. It is for this inescapable "need for exchange" that the brain builds, starting from perception, its most appropriate (for the purpose of this exchange) vision of the world, *its* Double. The *imagination* of what will happen in the world, on which the action is to be planned, is therefore of the Double. The action is of the brain; the prediction, linked to perception, is of the Double. The Double is the mind, inseparably linked to the brain, although functionally distinguishable from it.

The possibility of retro-action activity is observable in the laboratory in the formation of domains (or assemblies, in the jargon of neuroscience) of neurons oscillating in unison, modulated in amplitude and coherent in phase. The formation of these domains spreads not only from an apex, in the form of expanding cones, but also as cones that contract, converging to an apex. In the dissipative model these cones are described by divergent and convergent solutions, formally forward and backward in time (time-reversal), respectively. Laboratory observations give thus to the Double a structural consistency at the neuronal level (Freeman and Vitiello 2010, 2016; Freeman et al. 2012).

The dissipative model, by describing the dynamics of the processes that link the activity of the cortical mass of neurons (neuropil) to mental activity, thus suggests a possible solution to the "hard problem" (Atmanspacher 2015) of filling the gap between the strictly neuronal activity and the "qualia," the subjective mental experiences.

These processes underlying the construction of knowledge are entirely sustained by the perceptive experience and the formulation of hypotheses through mental activity (intentional imagination). This last one, only apparently separated from the neuronal matter, is based on the perception and is aimed to model "intentionally" the world (intentional perception). The credibility of the hypotheses and the vision so constructed rest on their verifiability through the "action," that makes the arrow of time and the causal sequence simultaneously perceptible.

The knowledge thus constructed gives us access to the world around us, gives it meaning, takes away space from the opacity that separates us from it.

References

Amaldi, E. (1962). Fisica Generale, Parte II, cap. X. In *La luce, Corpi trasparenti e corpi opachi* (p. 497). Roma: Tipo-Litografia Marves.

Atmanspacher, H. (2015). Quantum approaches to consciousness. In *Stanford encyclopedia of philosophy*. Retrieved from http://plato.stanford.edu/entries/qt-consciousness/

Cassirer, E. (1920). Das Erkenntnisproblem in der Philosophie und Wissenschaft der neueren Zeit, Dritter Band, Die Nachkantischen Systeme. Bruno Cassirer, Berlin, 4. "Die Zerlegung der Wirklichkeit in substantielle Einzelelemente, die selbständig für sich existieren und wechselweise in verschiedenartige Zusammenhänge des Wirkens eintreten, gilt demnach hier als ein fertiger Tatbestand, der vor aller näheren »erkenntniskritischen« Untersuchung als feststehend angenommen wird. Die Abstraktion, auf die sich die Sprache des Lebens und die Sprache der empirischen Wissenschaften gründet, mag noch so weit gehen: sie läßt doch stets gleichsam das kategoriale Grundgefüge der naiven Weltansicht unangetastet".

Celeghini, E., Rasetti, M., & Vitiello, G. (1992). Quantum dissipation. *Annals of Physics, 215*, 156–170.

Desideri, F. (2011). *La percezione riflessa*. Milano: Raffaello Cortina Editore.

Desideri, F. (2018). *Origine dell'estetico. Dalle emozioni al giudizio*. Frecce, Roma: Carocci Editore.

Freeman, W. J. (1975/2004). *Mass action in the nervous system*. New York: Academic.

Freeman, W. J., & Vitiello, G. (2010). Vortices in brain waves. *International Journal of Modern Physics B, 23*, 3269–3295.

Freeman, W. J., & Vitiello, G. (2016). Matter and mind are entangled in two streams of images guiding behavior and informing the subject through awareness. *Mind and Matter, 14*(1), 7–24.

Freeman, W. J., Livi, R., Obinata, M., & Vitiello, G. (2012). Cortical phase transitions, non-equilibrium thermodynamics and the time-dependent Ginzburg-Landau equation. *International Journal of Modern Physics B, 26*, 1250035. 29 p.

Merleau-Ponty, M. (1942/1963). *The structure of behavior* (A. L. Fischer, Trans.). Boston: Beacon.

Rossi, B. (1957). *Optics* (p. 404). Reading: Addison-Wesley Publishing Company, Inc.

Titus Lucretius Carus (99–55 B.C.). (2003). *De rerum natura. Liber I* (vv. 127–131, W. Englert, Trans.). Newbury Port: Focus.

Vitiello, G. (1995). Dissipation and memory capacity in the quantum brain model. *International Journal of Modern Physics B, B9*, 973–989.

Vitiello, G. (2001). *My double unveiled*. Amsterdam: John Benjamins Publ. Co.

Vitiello, G. (2004). The dissipative brain. In G. Globus, K. H. Pribram, & G. Vitiello (Eds.), *Brain and being. At the boundary between science, philosophy, language and arts* (pp. 315–334). Amsterdam: John Benjamins Publ. Co.

Vitiello, G. (2008). Essere nel mondo: Io e il mio Doppio. *Atque, 5*, 155–176.

Vitiello, G. (2016a). Opacità del mondo e conoscenza. *Atque, 8*, 17–32.

Vitiello, G. (2016b). Il meccanismo di Higgs e la freccia del tempo. *Rivista di Filosofia Neo-Scolastica, 2*, 361–371.

Vitiello, G. (2016c). …and Kronos ate his sons. In I. Licata (Ed.), *Quantum physics and spacetime dynamics beyond peaceful coexistence* (pp. 465–486). London: Imperial College Press.

References

In Search of Organization Laws: A New Way of Doing Science? (The Uprising of Systemic Attitude)

Alessandro Giuliani

Abstract The ever increasing need (especially in biology) to cope with problems asking for a network-like formalization in which the focus is a sensible description of the correlation structure among the constituent parts is catalyzing what appears as a change in the style of doing science.

The blurring of the distinction between hypothesis generating and testing processes and the substitution of theories peculiar of the specific investigation field with largely independent of the microscopic details organization principles are reshaping the scientific culture.

The above sketched style still refers to a minority of scientific works; nevertheless, it embeds a great promise of making science to exit the actual lack of efficacy crisis due to hyper-specialization.

1 An Information Crisis?

In 2005, a paper with the provocative title "*Why most published research findings are false*" (Ioannidis 2005) by the US-based Greek statistician John Ioannidis was like a rock falling in the pond of biomedical sciences. The initial reactions from scientific community ranged from denial to enthusiastic appreciation passing by a "moralistic-social" interpretation of the crisis as provoked by the misconduct of scientists that in turn was fostered by the "publish-or-perish" curse.

After more than 10 years from the publication of Ioannidis paper, it is now evident that moral and social dimensions had an irrelevant influence on the crisis: this is a real information crisis due to both the inadequacy of the great majority of biomedical scientists to grasp the meaning of statistical approach (Nuzzo 2014) (that in turn prompted the American Statistical Association to produce a document restating the fundamentals and epistemological status of the application of statistics on

A. Giuliani (✉)
Istituto Superiore di Sanità, Rome, Italy
e-mail: alessandro.giuliani@iss.it

L. Urbani Ulivi (ed.), *The Systemic Turn in Human and Natural Sciences*, Contemporary Systems Thinking, https://doi.org/10.1007/978-3-030-00725-6_3

empirical research (Wasserstein and Lazar 2016)) and to the positioning of the great majority of biomedical research at an inadequate (too detailed) noise-dominated level of investigation (Transtrum et al. 2015).

In the following, I will sketch some fundamental trends, already present in nowadays biomedical sciences, that hold promises to overcome the information crisis and to set the frame for a renewed research style.

2 Systemic Style

In his fundamental paper appeared in 1948 entitled "*Science and Complexity*" Warren Weaver (1948), one of the fathers of modern information science, proposed a tri-partition of science styles into: (1) Problems of simplicity, (2) Problems of disorganized complexity, and (3) Problems of organized complexity.

The first class (simplicity) collected all those problems that can be faced in terms of differential equations and thus well suited for deriving "general laws of nature." These "simple problems" were the ones solved by most "sophisticated" mathematics because they are amenable to a high degree of abstraction (e.g., a planet could be considered an abstract dimensionless "material point").

Problems of disorganized complexity (class 2) allow to get a still superior precision (and, most important to a higher degree of generalization than class 1) problems. These problems imply a somewhat opposite style of reasoning with respect to the "problems of simplicity." In this case, the efficiency does not stem from the possibility to get an abstract description of the involved players but from totally discarding such "atomic" knowledge in favor of very coarse grain macroscopic descriptors corresponding to gross averages on a transfinite number of atomic elements.

Both the above two methods meet drastic limitations of their applicability range. Class 1 problems need the presence of very few involved players interacting in a stable way with a practically null effect of boundary conditions, class 2 problems ask for very large number of particles with only negligible (or very stable and invariant) interactions among them.

Problems of organized complexity (Weaver class 3) arise in all those situations in which many (even if not so many as in class 2) elements are involved with non-negligible (and often time-varying) interactions among them, and with no possibility to sketch dynamical laws due to their extreme context dependence. This is the "middle kingdom" where life sciences live.

Before going ahead, it is worth reporting the original figure of the Weaver paper sketching the three realms of science (Fig. 1):

The left panel links are few and are unique for any couple of elements so allowing for a clear mathematical modeling. The right panel relative to organized complexity is the only proper "network": multiple links connect the elements; multiple equivalent paths can be used to explore the system.

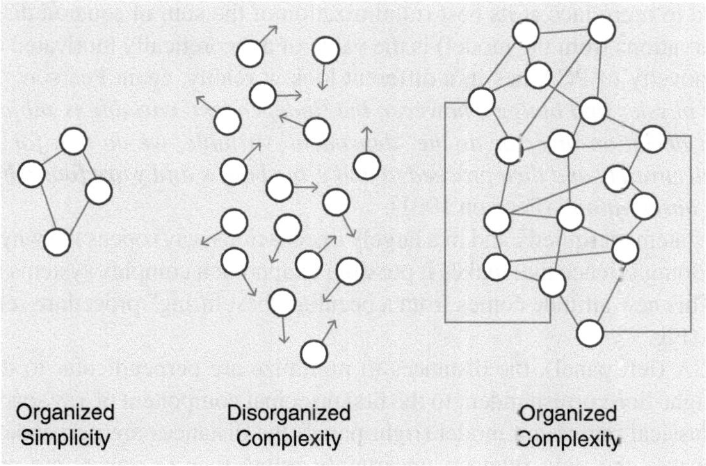

Fig. 1 The circles represent the elementary players, the lines their mutual relations. The lines of the graph in the middle (disorganized complexity) only symbolize the trajectories of the particles whose interactions are both random and contingent being limited to a huge number of random hits whose cumulative effect can be easily described in statistical terms

The main message Warren Weaver paper conveys can be summarized into: "*when dealing with complex systems, the focus of the investigation must shift from the detailed analysis of single elements to their wiring pattern.*" In 1948, Weaver was skeptical about the rise of a quantitative science of organized complexity, now we are in a more favorable (and optimist) situation.

In his seminal 1901 paper (Pearson 1901), Karl Pearson synthetically explained the motivation of developing Principal Component Analysis (PCA): "*In many physical, statistical and biological investigations it is desirable to represent a system of points in plane, three or higher dimensioned space by the 'best fitting' straight line or plane.*"

After this first statement, Pearson clarifies the novelty of his proposal with respect to the usual regression techniques: "*In nearly all the cases dealt with in the text-books of least squares, the variables on the right of our equations are treated as independent, those on the left as dependent variables.*" This implies that the minimization of the sum of squared distances only deals with the dependent (y) variable. The variance along independent (x) variable, being the consequence of the choice of the scientist (e.g., dose, time of observation) is supposed to be strictly controlled and thus does not enter in the evaluation of the "fit" of the model. in statistical jargon X is a "degenerate variable" totally free of errors.

This is the classical Galilean way to scientific enquiry: the scientist sets his/her experimental frame in terms of independent variable(s) (e.g., doses of drug to be tested, observation times, applied forces), and then he/she checks the degree of consistence of the observed values of a dependent (y) variable in terms of a pre-defined model. The uncertainty (errors, variability due to causes other than the x and so forth) only affects the dependent variable. The "reality" the fitting procedure is supposed to reproduce at its best (minimization of the sum of squared distances of the observations from the model) is the value of a theoretically motivated quantity.

The novelty of PCA lies in a different look at reality, again Pearson: "*In many cases of physics and biology, however, the 'independent' variable is subject to just as much deviation or error as the 'dependent' variable, we do not, for example, know x accurately and then proceed to find y, but* ***both x and y are found by experiment or observation***" (Pearson 1901).

This statement (quietly and in a largely unconscious way) opens the way to a new style of doing science that makes it possible to approach complex systems (Giuliani 2017). This new attitude comes from a peculiar "best fitting" procedure set forth by Pearson (Fig. 2).

In PCA (left panel), the distances to minimize are perpendicular to the model (the straight-line correspondent to the first principal component of x, y space), while in the classical regression model (right panel) the distances are perpendicular to x axis, because the only relevant uncertainty refers to y (x values are chosen by researcher). This apparently minor geometrical detail encompasses an epistemic revolution: the "real thing" is no more an observable motivated by an a priori theory like in classical regression, but a "hidden latent variable" emerging from the correlations actually observed in the analyzed data set.

The "hypothesis generating" and "hypothesis testing" phases normally separated in classical scientific method, conflate. A priori defined experimental observables are considered as variously biased versions of the latent reality that emerges as a "consensus axis" among different observables, the distances to minimize refer to a common "flux of variation" of two (or more) experimental observables.

The implicit paradigm is that there is a "reality behind the curtain," that shapes the mutual correlation among the observables: the actual measures are not the real (even if blurred by the measurement errors) thing, but only the image in light of something else (Giuliani 2017).

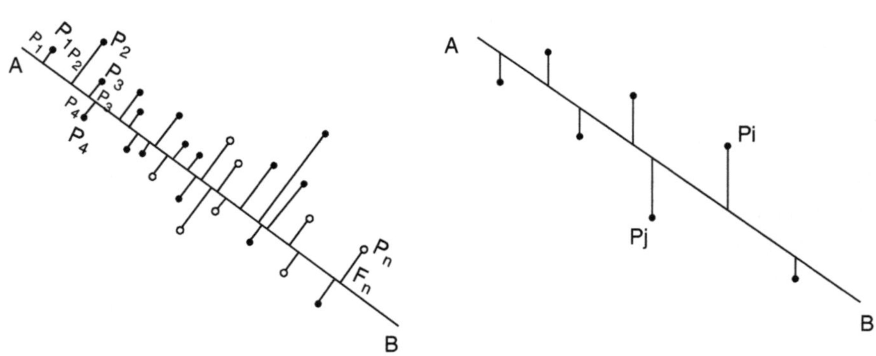

Fig. 2 In the left panel (PCA), the least square optimization builds upon the actual distances of experimental points in the space. In the right panel (usual regression), only the variability along Y axis is taken into consideration, X axis variability is "degenerate" (i.e., it only depends on experimenter's choice)

Both Weaver theoretical and Pearson mainly operational (PCA is the by far most widespread multidimensional data analysis tool) proposals redound around the same issue: *when describing phenomena involving many different and inter-related players we can no more testing our hypotheses on an element-by-element basis, we must focus on relation structure and let the latent variables implicit in the wiring structure to emerge. This is the essential of what we are used to call "systemic style"* (Huang 2004).

3 Networks

As aptly pointed out by Nicosia et al. (2014) "*Networks are the fabric of complex systems*," we already pointed out the centrality of the concept of network for organized complexity, this is why apparently disparate investigation fields—from protein science (Di Paola et al. 2012) to psychiatry (Hauser et al. 2016)—are now starting to share the principle that natural sciences must look for a unification frame not by investigating the deep structure of matter, but by exploiting the phenomenological consequences of shared organization rules governing the relations (interactions, correlations,) between the constitutive elements of the system at hand.

As a matter of fact (Laughlin et al. 2000), Laughlin and colleagues identify the frontiers of basic science in "*the search for the existence and universality of such rules, the proof or disproof of organizing principles appropriate to the mesoscopic domain.*"

The quest for "network laws" only stemming from wiring architectures and largely independent of the nature of the constituting nodes of the network, stems from the work of the Dutch electrical engineer Bernard Tellegen (1952) that, in 1952 developed a sort of conservation principle (tailored upon Kirchoff's laws of electrical circuits) of both potential and flux across a network. The flux does not need to be an electrical current and the same holds for the potential. Any system that can be modeled by a set of nodes linked by edges (being them metabolites linked by chemical reactions transforming one into the other or mutually interacting persons in an office) has similar emerging properties independently of the physical nature of nodes and edges. As aptly stressed in (Mickulecki 2001), the theorem opens the way to a sort of "network thermodynamics," whose principles are strictly dependent from wiring architecture while largely independent of the constitutive laws governing the single elements.

Network graph-theoretical approaches (a mathematical graph is fully equivalent to a network expressed in terms of its adjacency matrix) are located half-way between bottom-up and top-down approaches focusing on the relation between the elements of the studied phenomenon. We can roughly describe the network approach as the answer to the question "*What can we derive from the sole knowledge of the wiring diagram of a system?*"

Graphs are described by measurements located at local (single nodes), global (entire network), and mesoscale (clusters of nodes, optimal paths) levels. Thus, we can compute the degree of each node (how many links are attached to a given node) that is a local descriptor, or we can compute the so-called average shortest path or characteristic length of a graph corresponding to the average length of minimal paths connecting all the node pairs (this corresponds to a mesoscopic feature of the system).

Eventually, we can compute a global feature like the general connectivity of the network (density of links) (Csermely et al. 2013).

Figure 3 reports an exemplar network structure with the indication of some relevant network invariants, structural descriptors of the wiring architecture: each descriptor can be referred to a single node (microscopic level) but its value depends on the position of the node within the network, in the same way the descriptors computed at the entire network level stem from averages on the single nodes. This in some way creates a "natural" microscopic-macroscopic link devoid of any strong theoretical assumption.

The different organization layers are strictly intermingled and cannot be decoupled: they derive from the same basic representation (the graph) and any view influences (and it is influenced by) all the others. In other words, it does not exist a unique "privileged layer" where "the interesting facts" happen. The role of a node strictly depends upon its position in the network (top-down causation), while the global properties of the network strictly rely on the single nodes wiring patterns (bottom-up causation).

We refer to this kind of global organization as "middle-out" (Giuliani et al. 2014) to stress the fact that the uncovering of the mutual relation among the parts represents the core of the explanation from where to start to grasp the entire frame.

We neither go "top-down" (general laws dictate the behavior of specific cases like in problems of organized simplicity where, as for gravity, a cat is fully equivalent

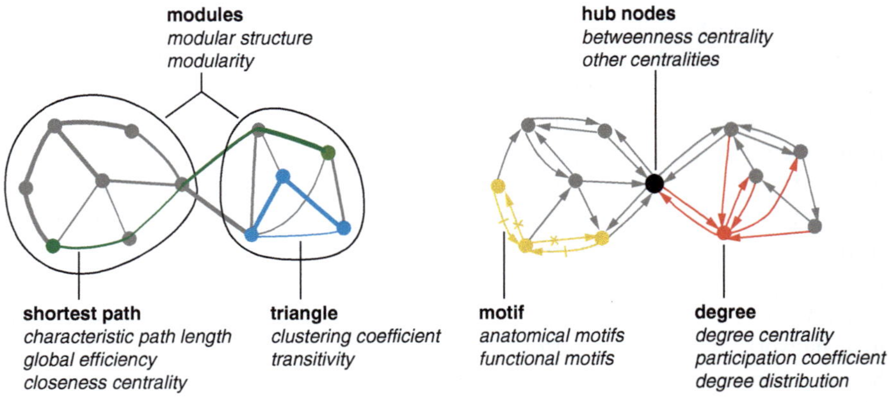

Fig. 3 Modules correspond to subset of nodes having much more links among them than with other nodes of the network. Measures of centrality (closeness, betweenness...) describe nodes in terms of the number of shortest paths traversing them. Shortest path is the characteristic metrics for networks: they correspond to the shortest distances (in terms of number of nodes/links to be traversed) for linking pairs of nodes

to a chair) nor "bottom-up" (statistics over large ensembles promulgate the laws governing the entire system like in disorganized complexity cases).

Approaching a problem by a "network-based approach" implies the collapse of a huge set of "microscopic information" in the form of "who-is-connected-with-whom" (think of the astronomic value reached by the $N^*(N - 1)/2$ distinct possible pairwise relations between different gene expression values in the case of a classical microarray experiment when N is around 30,000! that, if one-by-one exploited, should give rise to a plethora of chance correlations) into coarse grain statements like "*The treatment X drastically increases the average shortest path of the system with a consequent decay of signaling efficiency*" or "*The percentage of variation explained by first principal component decreases steadily with the gravity of the disease Y so pointing to a progressive loss of connectivity of the underlying network.*"

In order such statements could be of use for biomedical scientists, there is the need of a cultural change that urges scientists from different disciplines to find a "shared playground" made of common concepts from statistical mechanics but largely devoid of mathematical formalism, it is mandatory to have a general appreciation of the definition of order and organization in terms of correlation structure (see (Gorban et al. 2010) for a very brilliant example) and a capacity to look at the specific field of interest without eliminating the "big picture" in which is embedded into, so to find the proper formalization of the problem.

4 Conclusions

It is sufficient to interrogate a scientific literature repository with statements like "*complex networks gene expression*" (15,700 results since 2014 in Google Scholar the 4th June 2018) or "*multidimensional statistics neuroscience*" (17,600 results since 2014 in Google Scholar the 4th June 2018) to have a glimpse of a mounting wave of scientific work following the general trend of "focusing on relations" and thus of the rising number of scientists (more or less consciously) acquiring a systemic attitude. If and when such a trend will give birth to a new style of doing science that gradually will escape the bottle-neck of the ultra-reductionist approach promoted by molecular genetics so to establish a sort of "biological statistical mechanics" is impossible to predict. What is for sure is that biomedical sciences are actively reshaping and this process will have deep cultural consequences.

References

Csermely, P., et al. (2013). Structure and dynamics of molecular networks: A novel paradigm of drug discovery: A comprehensive review. *Pharmacology & Therapeutics, 138*, 333–408.

Di Paola, L., De Ruvo, M., Paci, P., Santoni, D., & Giuliani, A. (2012). Protein contact networks: An emerging paradigm in chemistry. *Chemical Reviews, 113*(3), 1598–1613.

Giuliani, A. (2017). The application of principal component analysis to drug discovery and biomedical data. *Drug Discovery Today, 22*(7), 1069–1076.

Giuliani, A., et al. (2014). Why network approach can promote a new way of thinking in biology. *Frontiers in Genetics, 5*(83), 1. https://doi.org/10.3389/fgene.2014.00083.

Gorban, A. N., Smirnova, E. V., & Tyukina, T. A. (2010). Correlations, risk and crisis: From physiology to finance. *Physica A: Statistical Mechanics and its Applications, 389*(16), 3193–3217.

Hauser, T. U., Fiore, V. G., Moutoussis, M., & Dolan, R. J. (2016). Computational psychiatry of ADHD: Neural gain impairments across Marrian levels of analysis. *Trends in Neurosciences, 39*(2), 63–73.

Huang, S. (2004). Back to the biology in systems biology: What can we learn from biomolecular networks? *Briefings in Functional Genomics, 2*(4), 279–297.

Ioannidis, J. P. (2005). Why most published research findings are false. *PLoS Medicine, 2*(8), e124.

Laughlin, R. B., Pines, D., Schmalian, J., Stojković, B. P., & Wolynes, P. (2000). The middle way. *Proceedings of the National Academy of Sciences, 97*(1), 32–37.

Mickulecki, D. (2001). Network thermodynamics and complexity: A transition to relational systems theory. *Computers and Chemistry, 25*, 369–391.

Nicosia, V., De Domenico, M., & Latora, V. (2014). Characteristic exponents of complex networks. *EPL (Europhysics Letters), 106*(5), 58005.

Nuzzo, R. (2014). Scientific method: Statistical errors. *Nature News, 506*(7487), 150.

Pearson, K. (1901). On lines and planes of closest fit to systems of points in space. *The London, Edinburgh, and Dublin Philosophical Magazine and Journal of Science, 2*(11), 559–572.

Tellegen, B. (1952). A general network theorem with application. *Philips Research Reports, 7*, 259–269.

Transtrum, M. K., Machta, B. B., Brown, K. S., Daniels, B. C., Myers, C. R., & Sethna, J. P. (2015). Perspective: Sloppiness and emergent theories in physics, biology, and beyond. *The Journal of Chemical Physics, 143*(1), 07B201_1.

Wasserstein, R. L., & Lazar, N. A. (2016). The ASA's statement on p-values: Context, process, and purpose. *The American Statistician, 70*(2), 129–133.

Weaver, W. (1948). Science and complexity. *American Scientist, 36*, 536–549.

Is Present Ecology a Systemic Discipline? New Scientific Paradigms Lead to Bionomics

Vittorio Ingegnoli

Abstract If the basilar concept of Ecology is the "Ecosystem" is this a desecrating title? The answer depends on considerations related to the challenges come up by Reality, which is complex, creative and needing freedom: these questions brought Galilean scientific method and Science to crisis and drove them towards a change of its fundamental principles, therefore to new scientific paradigms. That's the reason of new biological disciplines, more available to study complex systems, e.g. epigenetics, agroecology, systemic medicine, bionomics, and environmental health. All of them, especially bionomics, underline the limits of ecology in studying complex systems: here the *ambiguity* of the concept of *ecosystem*, a reinterpretation of *biodiversity* and *resilience* and the completion of the *Spectrum of Biological Organization* on Earth are briefly enlightened.

Well, what's Bionomics? It's the new discipline investigating the *Laws of Life on Earth* as a hierarchical organization of complex systems, acting as living entities: so, it transforms many principles of traditional Ecology and confirms the preeminent importance of the systemic approach to correctly evaluate and care the Planetary Health, to which it gives a wide theoretical corpus. A short synthesis of some of the main aspects of Bionomics and Landscape Bionomics is given.

1 Going Over the Galilean Method

The necessity of a scientific *paradigm shift* is today impellent and emerges from many questions in many scientific fields. The radical empiricism is linked with determinism, which gives no importance to time and put in question the human freedom and the knowledge of the real world: so in a world dominated by necessity there is no place for creativity.

V. Ingegnoli (✉)
Department of Environmental Science and Policy—ESP, Università Statale di Milano, Milan, Italy
e-mail: vittorio.ingegnoli@guest.unimi.it

L. Urbani Ulivi (ed.), *The Systemic Turn in Human and Natural Sciences*, Contemporary Systems Thinking, https://doi.org/10.1007/978-3-030-00725-6_4

Many epistemological studies indicate that the modern scientific method, i.e. Galilean, deductive, experimental, although not yet been passed, has shown severe limitations. Thus, what are the limitations of the Galilean method, the basis of modern science today no more suitable? Remember that this fundamental start of modern natural science consciously proceeded with a cognitive restriction: it considered modern natural science as something different from a metaphysic of nature, adding also the restriction of the necessity of a quantification by mathematics. So, as well expressed by Agazzi (2014), the elimination of the *final causes* or *intrinsic goals* of natural entities from natural sciences appears as an arbitrary limitation in the study of actions and of living organisms too.

Another limitation of the Galilean method concerns the meaning of information. Since a processor can interpret messages only if it shares an encoding, a functional protein must be compatible with its interpreter, hence it cannot be generated randomly: if not, in system with high complexity, it can only lead to useless or destructive changes. The coding has also to do with the meaning of information: putting together a series of terms—or dialling randomly some sentences (if that were possible) using the words founded in a dictionary—is not enough if you want to write a poem.

Note that the use of the term *information* might be misleading, as it depends on the concept of *compressibility*. Informally, from the point of view of Algorithmic Information Theory (AIT), the information content of a string is equivalent to the length of the shortest possible self-contained representation of that string. A self-contained representation is essentially a programme. Following the AIT the complex specified information (CSI) is the *incompressible* information, which cannot be synthesized in a more simple form or rule. For example, let us consider a DNA sequence like this: CTAGGCATCATGAAATAGGAACAAATCATTTAG. No chemical or physical law contains this sequence or the description of the other organic macro-molecules. Being these sequences incompressible, they cannot be generated by natural laws which are simple algorithms shorter than a sequence.

Let's observe that the scientific knowledge constructed in each part of science, which is separated from other parts, cannot be united by science itself, as underlined by Agazzi (2014). To restore a sort of unity of knowledge, we must appeal to philosophy. Einstein wrote in his Scientific Autobiography (Einstein 1949) that even scholars of relief (such as Mach) could be hampered by prejudice in the interpretation of the facts: Prejudice, which still is not gone, is the belief that the facts can and should result in scientific knowledge by itself, without free conceptual construction. The same great scientist (Einstein 1944) adds that independence from the prejudices is determined by philosophical analysis and it's the mark of distinction between a mere artisan or specialist and a real seeker after truth.

The *study*, the research of the truth, needs a spirit of participation. The consonance, one of the attribute of love, what brings the unity among the contrasts, law of just balancing, means *harmony*. The *study* of *harmony* is the unique way to reach the comprehension (Ingegnoli 1980). The crucial importance of systemic epistemology derives from principles like these. Even the epistemologist Bronowski (1969) states that what will allow humanity to survive and to continue in scientific discoveries will not be just or unjust rules of conduct, but *more deep illuminations*, at the light of which good and evil, means and ends, justice and injustice will be seen in a terrible clarity of boundaries.

2 Limits Emerging from Paradigm Shift

2.1 Limits of Conventional Ecology

While the development of System Theory originated from Biology and while complex systems brought directly to change the scientific paradigm, the discipline of Biology remained anchored to Galilean method and to reductionism. Some examples:

(a) Hemato-encephalic barrier theory (1913), taught until few years ago, even if Besedovsky demonstrated on 1981 (Besedovsky et al. 1981) the direct interrelation immune system/brain.
(b) Neo-Darwinian theory, variation and natural selection, chance and necessity, with many supporters asserting it as *the* theory of evolution still today, ignoring or disregarding the scientific objections starting since Waddington (1942) (Waddington 1942, 1957) with epigenetics.
(c) The "central dogma of molecular biology" (Crick 1970) in which DNA is responsible of all the characters of an organism, with a direct passage genotype-phenotype, reinforcing the previous point.
(d) The big research on human genome, imposed by the central dogma, whose results helped to strongly limit this dogma.
(e) The concept of "ecosystem" (Tansley 1934) broadly and inaccurately used even today ignoring the implications in contrast with the System Theory, producing ambiguity.
(f) The "green revolution", linked to GMO, industrial agriculture, dominating agrarian science and applications since one century, whose excess of simplification of agricultural landscapes improved the environmental degradation of the Earth.

Undoubtedly, the current change of paradigm is much more hard than in the past shifting and presents much more difficulties, because after so many centuries it gives a stop to the *positive* evolution of Science, putting it in crisis. This is a good sign, because we have to underline that the reality is in itself capable to lead towards the truth and consequently the crisis of science will be surpassed. The entire humankind should receive a benefit, due to the universality of science. As confirmed by Agazzi (2014), the more advanced epistemology can be recognized by the *inversion* of the conventional neo-positivist steps: from science to metaphysics to religion. This fact gives the extraordinary dimension of present scientific paradigm shift.

Going back to the title of this chapter "*Is Present Ecology a Systemic Discipline?*", a similar question may seem desecrating: if the basilar concept of Ecology is the "Ecosystem", how dare you dare to ask yourself something like that? The answer depends on various considerations: here we can only underline few of its deep limits regarding the concept of ecosystem, resilience/resistance, biodiversity and show a consequence.

Remember that the concept of ecosystem was proposed by Tansley in the years 1930–1935 as the set of living organisms and inanimate substances exchanging materials and energy in a limited area. Later on, Odum (1971, 1983) specified the structure and functions of the ecosystem, noting that the term ecosystem may be individualized from a small temporary pond of few square metres to an entire Alpine valley of hundreds of square kilometres and of thousand years.

On 1986, a group of American ecologists (O'Neill et al. 1986) published a book demonstrating that the ecosystem is an ambiguous concept: a real natural formation (e.g. a small ecotope) can be studied with a Biotic Emphasis (community) or a Functional one (ecosystem), but the two analysis cannot be truly integrated (Fig. 1, left); Bailey (1996) (Fig. 1, right) underlined that a system of different ecosystems is named again ecosystem! It's not a question of name or of nested structure: here the principle of emergent properties is completely ignored! It's impossible to study complex systems using concepts contradicting the System Theory! That is why Ingegnoli (2001, 2002) criticized the ecosystem and defined the landscape in a systemic way (see Fig. 6 in Sect. 4). Moreover, some people (e.g. media) use the term ecosystem giving only a generic sense of "ecological system", but this has nothing to do with Science.

The limits of conventional ecology also reflect in the incorrect use of the concepts of resilience and biodiversity. As underlined by Odum (1971, 1983) and demonstrated by Pignatti (1995) on the Mediterranean Phytocoenosis, most evolved natural systems follow the laws of "resistance stability", not the ones of the "resilience stability", therefore the specific biodiversity in these case is low, as we can see in Fig. 2.

Why? The more ordered an ecological system is, the more its component reach new and efficient links capable to incorporate the disturbances and to defend the

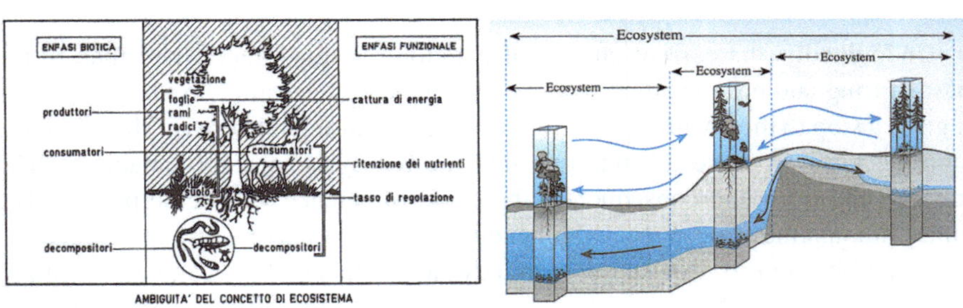

Fig. 1 The ambiguity of the ecosystem concept (O'Neill et al. 1986; Bailey 1996; Ingegnoli 2001, 2015) in which (left) the biotic viewpoint and the functional one can't be integrated and the emergent properties (right) are ignored

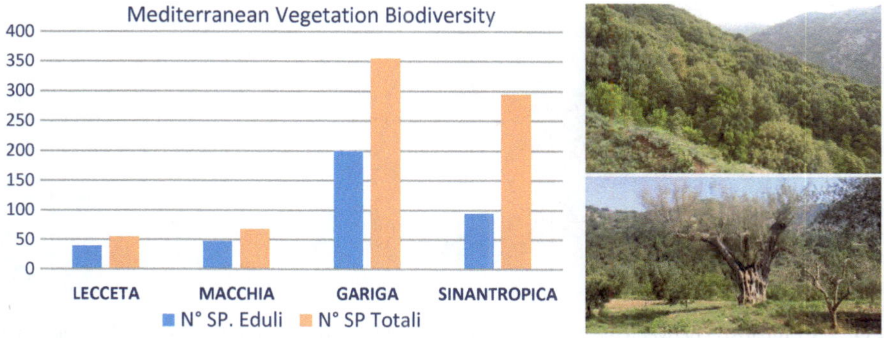

Fig. 2 The species biodiversity of Mediterranean vegetation. In oak forest (lecceta, photo above) we may find 40–50 species per survey, while in degraded areas (gariga) the species arrive to be more than 300. Photo below synanthropic vegetation

system by bigger perturbations. Only resilience needs a redundancy of species (more elevated biodiversity) to react through species replacement to the lack of organization. The resilience of natural systems is frequently a surviving strategy when the limits to disturbances incorporation are exceeded. For instance, after a perturbation capable to destroy a forest area, quickly growing herbaceous and shrub species are able to reduce the transformation deficit, waiting the trees regrown. It's the case of Fig. 2 passing from a sclerophyll forest of *Quercus ilex* (Lecceta) to a Gariga, that is from 40–50 species to 300–320.

No doubt that biodiversity cannot be destroyed without dangerous results implying the order of life systems: e.g. a Lecceta with only 12 species is not generally in a good state. Moreover, in a landscape both resistance and resilience subsystems are present, even if the strategy of nature privileges resistance: it is a survival reinforcement. In fact, let us observe that the very high specific biodiversity proper of tropical forests can be the reaction of the Earth to a disturbance due to the too hot climate of the present inter-glacial period. As affirmed by Lovelock (2007), so rich biodiversity is not necessarily highly desirable and to be preserved at all costs. Finally, biodiversity is mainly intended as related to species, forgiving the landscape biodiversity, related to ecotopes and landscape units, linked with systemic variables and parameters.

The limits of conventional ecology may produce errors in their applications, too. An example is the detection of the Metropolitan Area of Milan, named Milano Città Metropolitana (MCM), of recent administrative formation.

The MCM area has been delimited following the parameters of conventional ecology, using the concept of eco-mosaic, the districts boundaries, the geographic density of population and reductionist parameters. Results of a parallel metropolitan study (MMS), following systemic bionomic concepts, show (Fig. 3) the inclusion of the NW and North parts, i.e. the province of Monza-Brianza and the Southern part of the provinces of Como and Varese: actually Monza is the most urbanized area and traditionally the most linked with Milan. The Metropolitan Area of Milan in the first case (MCM) is inhabited by 3.2 million people, while in the other case (MMS)

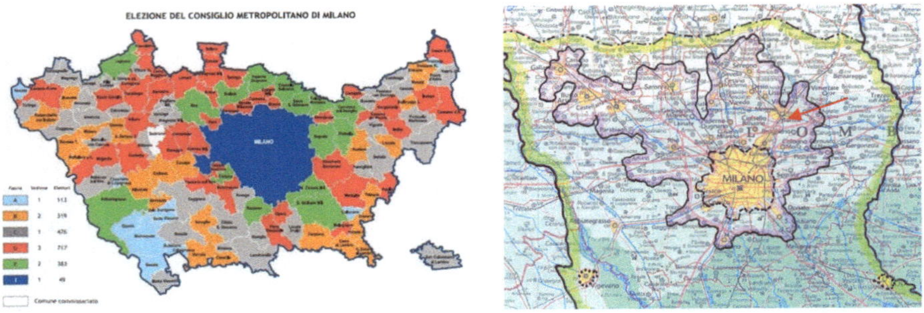

Fig. 3 Comparison between the Metropolitan Area of Milan identified following the Conventional Ecology (MCM, left) or the Landscape Bionomics (MMS, pink boundary, right)

it reaches 5.0 million. Moreover, MCM includes many agricultural municipalities which risk to become more and more urbanized. A right governance needs correct delimitations of a problem.

Limitations like these are very heavy and many scientists tried to update the conventional ecology. So, following new scientific paradigms, after a hard pioneer's effort, new biological disciplines emerge, the most important of which for our research can be: Agroecology, Environmental Health, Urban ecology, Bionomics and Landscape Bionomics (Fig. 4).

2.2 *Limits of Conventional Vegetation Analysis*

Concerning current methodologies of vegetation studies too, safe limitations are evident being phytosociology, sensu Braun Blanquet (1926), based on a reductionist epistemology and on ecological concepts today surpassed, even if have obvious advantages with regard to the description.

Here a quick overview of the main limitations of phytosociology (Naveh and Lieberman 1984; Pignatti et al. 2002; Ingegnoli 2002, 2011): (a) reference to a concept of naturalness that excludes humans in any event; (b) dynamics based on the concept of ecological succession mainly understood as linear and deterministic; (c) reference to an "ecological space" that does not consider the principle of emergent properties (Fig. 5); (d) use of the concept of "potential vegetation" not considering the role of disturbances in ecological systems; (e) ignorance of complex system and scale-dependent functions, claiming to be able to study the landscape with the deterministic approach of the concept of "sygmetum" and "geo-sygmetum" by Tüxen (1956) and Rivas-Martinez (1987).

A systemic review, consistent with the theory of landscape bionomics, starts with the proposal of the new concept of *the fittest vegetation for…*: this reinterpretation of the concept of *potential vegetation* indicates "the vegetation most fitting in climatic and geomorphic conditions, in a limited period of time, in a certain defined place, in function of the history of the same place and with a certain set of

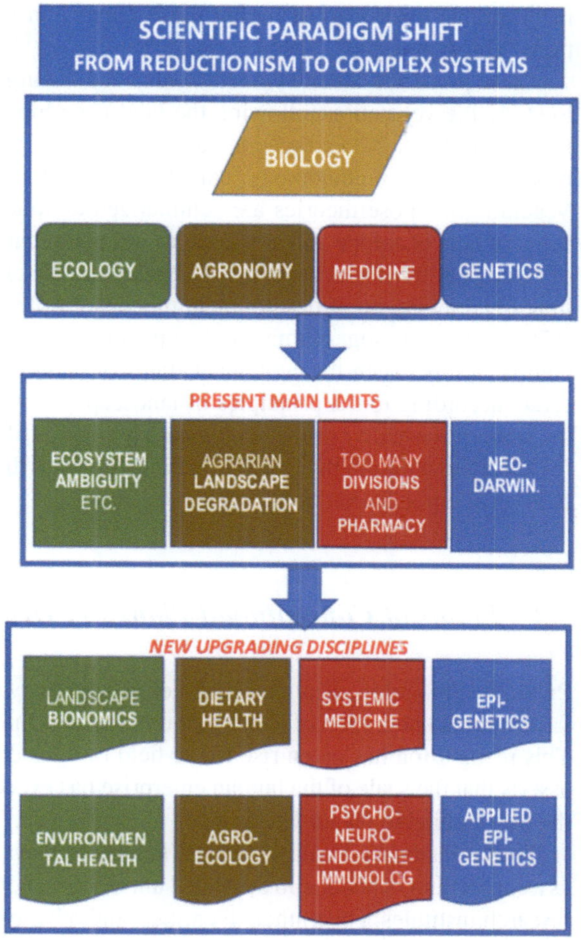

Fig. 4 The main biologic disciplines, born to better upgrading the scientific paradigm shift

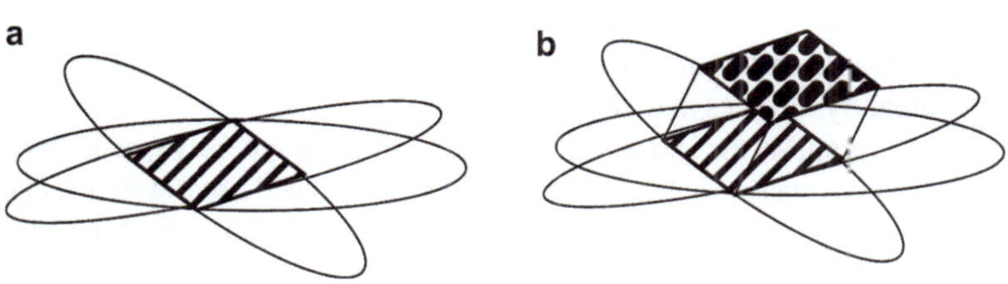

Fig. 5 Representation of an "ecological space" in the study of vegetation: (**a**) in the phytosociological model; (**b**) in the bionomics model of the landscape. Note that the principle of emergent properties acquires ecological space characters that go beyond the sum of individual species (Ingegnoli 2002)

incorporable disorders (including those human) in natural and not natural conditions" (Ingegnoli 2002, 2015).

These premises, albeit synthetic, enhanced the urgent need to develop a method to study the vegetation following the landscape bionomics principles. That's why Ingegnoli (2002, 2011), Ingegnoli and Giglio (2005), Ingegnoli and Pignatti (2007) proposed a new methodology, called LaBiSV (Landscape Bionomics Survey of Vegetation), whose theories are summarized as follows: (1) reference to the concepts of ecocoenotope and ecotissue as structural entities of the landscape; (2) use of biological territorial capacity of vegetation (BTC) as the main integrative function; (3) drawing up of development models of different types of vegetation (time-BTC) based on logarithmic and exponential functions; (4) the possibility of comparison between the ecological status of natural and man-made vegetated tesserae, according to the principles of landscape bionomics; (5) ability to determine the state of normality of the ecological parameters of different types of vegetation; (6) ability to measure the concept of biodiversity at the landscape level (diversity of biological organization of the context).

2.3 *Limits of Conventional Planetary Health Studies*

Public health and medical communities around the world are recognizing that pervasive human alteration of Earth's natural systems threatens the health of humanity. This recognition has given rise to the field of "planetary health": Planetary Health asserts that the scale of the human enterprise has exceeded available resources from the only habitable planet we know.

Thus a Planetary Health Alliance (PHA) has been founded in Harvard on 2016, as a consortium of about 100 dedicated universities, non-governmental organizations, research institutes, governmental entities, and other partners from 29 countries with a shared mission: supporting the growth of a rigorous, policy-focused, transdisciplinary field of applied research aimed at understanding and addressing the human health implications of accelerating environmental change, consequently based on the relations Geo-Health/Human-Health, that is the strictly relations between Geo-Environmental Syndromes and Human Diseases.

One could suppose the existence of a deep interchange between Ecology and Medicine: but here many problems arise. The PHA goals need a real systemic approach in the studies, because, as we will see, the upper organization levels of the biological spectrum (and the Earth itself) are complex systems, living entities, of one of which—the most politically involved, i.e. the landscape—man is the most peculiar component. Therefore, a living entity must be investigated in its physiology (e.g. metastability, biologic functions, autotrophy etc.) and pathology by a systemic advanced bio-ecological discipline, like Bionomics is. This is to check "if", "how" and "how much" environmental alterations could reflect on human health, independently from pollution.

On the contrary, present analyses are still mainly reductionist and more centred on medical disciplines than on a balance between ecology and medicine. Thus, the considered alteration of the environment leading to alteration of human health is only due to: (a) climate change, (b) resource scarcity, (c) land use change, (d) altered biogeochemical cycles, (e) biodiversity loss, (f) global pollution. Systemic alterations are *not mentioned at all*. Consequently, another important group of dysfunctions is forgotten: (g) Living complex systems dysfunctions (e.g. landscape unit alterations). Note that (g) is the real subject of all ecological determinants, the pivot of their complex system network.

Again, we can see that, to reach the mentioned balance ecology–medicine, it is absolutely indispensable to upgrade the today conventional ecology, with the help of bionomics.

3 Bionomics and Landscape Bionomics

We have to start observing that life is a complex self-organizing system, operating with continuous exchange of matter and energy with the outside; the system is able to perceive, to process and transfer information, to follow rules of correspondence among independent worlds (coding), to reach a target, to reproduce itself, to have an history and to participate in the process of evolution. Moreover, we observe that, in an evolutionary view, structure and function become complementary aspects of the same evolving whole.

Consequently life cannot exist without its environment: both are the necessary components of the system, because life depends on exchange of matter and energy and information between a concrete entity, like an organism or a community, and its environment (Ingegnoli and Giglio 2005). That is the reason why the concept of life is not limited to a single organism or to a group of species and therefore life organization can be described in hierarchic levels (i.e. the so-called "biological spectrum" sensu Odum (1971, 1983)). The world around life is made also by life itself; so the integration reaches again new levels. This is the reason why biological levels cannot be limited to organism, population, communities and their life support systems: as clarified by Bionomics, life includes upper scale complex ecological systems such as ecocoenotopes[1] (Ingegnoli 2002), landscapes[2] (Ingegnoli 2011), ecoregions (Bailey 1996) and the entire ecosphere (ecogeobionoosphere) too. As all remember, the Gaia Theory (Lovelock 2007; Lovelock and Margulis 1974) has already asserted that the Earth itself is a living entity.

So, the new discipline of Bionomics, derived from pioneer studies of Ingegnoli (1971, 1980, 1991) and discussions together with Richard Forman (Harvard), Zev Naveh (Haifa) and Sandro Pignatti (Rome), radically transforms the main principles of traditional Ecology by being aware that Life on Earth is organized in hierarchical levels, each one of them being a type of complex system, better a really existing living entity: again a sharp difference with the reductionist approaches to the study of the environment, through four parallel not-interacting hierarchies, respectively

[1] The *ecocoenotope* is the ecobiota, composed by the community, the ecosystem and the microchore (i.e. the spatial contiguity characters, sensu Zonneveld (1995: 51–59)).

[2] The *landscape* is a complex system of interacting ecocoenotopes (the "green row" in Fig. 6).

Hierarchic levels of Biological Organisation on the Earth

Scale	Viewpoints				REAL SYSTEMS[5]
	SPACE[1] CONFIGURATION	BIOTIC[2]	FUNCTIONAL[3]	CULTURAL-ECONOMIC[4]	
Global	Geosphere	Biosphere	Ecosphere	Noosphere	Eco-bio-geo-noosphere
Regional	Macro-chore	Biome	Biogeographic system	Regional Human Characters	Ecoregion
Territorial	***Chore***	***Set of communities***	***Set of Ecosystems***	***District Human Characters***	***Landscape***
Local	*Micro-chore*	*Community*	*Ecosystems*	*Local Human Activities*	*Ecocoenotope*
Stationary	Habitat	Population	Population niche	Cultural/Economic	Meta-population
Singular	Living space	Organism	Organism niche	Cultural agent	Meta-organism

1= not only a topographic criterion, but also a systemic one; 2= Biological and general-ecological criterion; 3= Traditional ecological criterion; 4= Cultural intended as a synthesis of anthropic signs and elements; 5= Types of living entities really existing on the Earth as spatio-temporal-information proper levels

Fig. 6 The Hierarchy of Life Organization on Earth following Bionomics

biotic, functional, spatial/configurationally and cultural/economics, all recognizing only six levels of scale (Fig. 6).

No doubt that some characters of community and ecosystem are available also at landscape level and even the inverse is true: only reductionism pretends to separate all the characters related to each level. For example, processes allowing the definition of life are exportable characters: each specific biological level expresses a process in a proper way, depending on its scale, structure, functions, amount of information and semiology. But we can note that each biological level presents exportable characters and proper ones (Ingegnoli 2002): as each system which owns proper characters is an entity, and we can find emergent properties characterizing each one of the previous levels, they are six types of concrete living entities, whose investigation needs these criteria to be reconceived, remembering that any ecological system must include both a biological element and its environment, plus its cultural/information contents.

The System Theory affirms that the scale capable to maximize the importance and the quantity of relations among the components of a system is the scale which consents to discriminate the different forms, especially the relational ones. That's why Ingegnoli (2002) and Ingegnoli and Giglio (2005) enhanced the crucial importance of the landscape, because territorial is the best scale capable to maximize the importance of the relations among the elements, both natural and human.

So deepen our understanding of "Biological-Integrated Landscape Ecology" (Ingegnoli 2002), the "Landscape Bionomics" arises (Ingegnoli 2011, 2015). The attempt to understand the behaviour of a landscape elaborating its thematic components meta-data (i.e. species, soils, human activities, hydrology, etc.), even with the help of GIS mapping and statistic controls or computer clustering landscape indicators (to be supervised and strictly limited), or scaling up an ecological system of communities, is without hope. The principle of emergent properties demonstrated the necessity of a top-down main criterion of observation to enlighten and preserve the new acquired systemic properties.

A synthetic presentation of the main principles and methods of landscape bionomics will confirm the intrinsic systemic criteria of this new discipline.

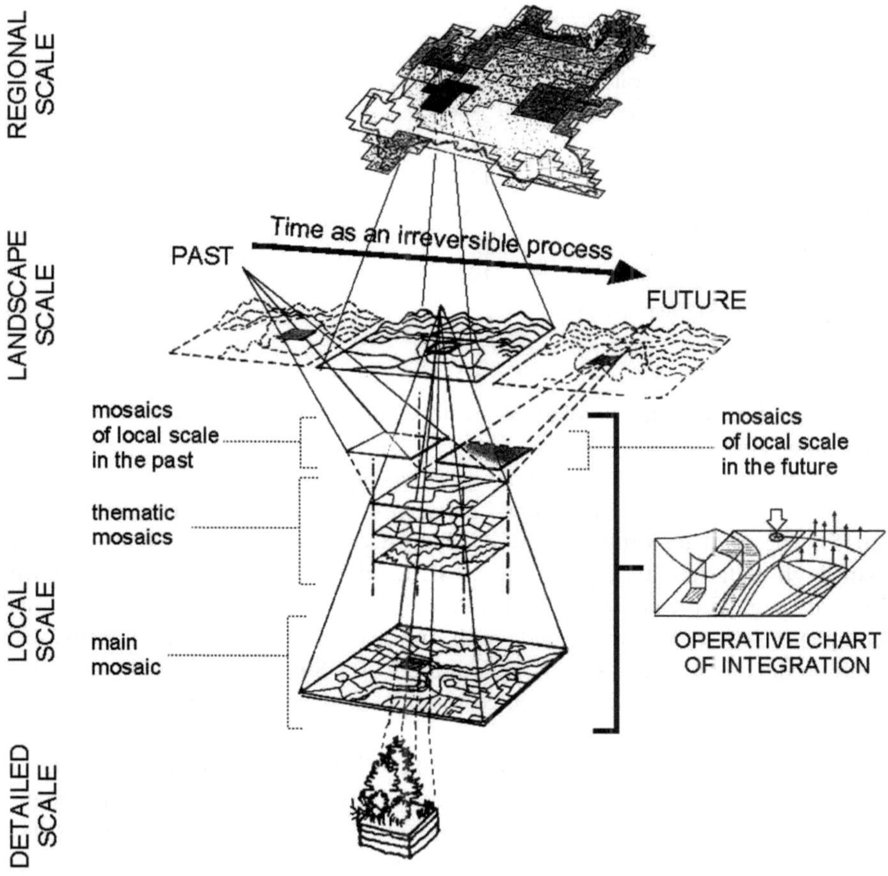

Fig. 7 The concept of ecotissue (Ingegnoli 2002, 2011, 2015). The basilar eco-mosaic is referred to vegetation. The complex structure of a landscape derives from the integration of different components: temporal, spatial, thematic

3.1 Landscape Structure (Shape–Function Relation)

The fundamental structure of a landscape is clearly systemic, an ecological tissue as the weft and the warp in weaving or the cells in a histologic tissue. The concept of *Ecotissue (Ects)* concerns a multidimensional conceptual structure representing the hierarchical intertwining, in past, present and future, of the ecological upper and lower[3] biological levels and of their relationships in the landscape (Giglio 2002), represented by a basic mosaic and a hierarchic succession of correlated mosaics and attributes (Fig. 7).

The *Landscape unit (LU),* intended as a sub-landscape, is a part of a landscape which assumes particular characters or even functions in relationship to the entire

[3] Ecoregions and ecocoenotopes, definitions in Giglio (2002: 323–333).

Table 1 Theoretical minimum standard habitat/capita

Climatic belts	Needed		
	Kcal/inhab[a]	SH* m²/capita	Agricultural surface/capita
Arctic	3500	2500	1670
Boreal	3100	1850	1250
Cold-temperate	2850	1480	1050
Warm-temperate	2750	1360	980
Sub-tropical	2550	1250	870
Tropical	2350	1020	730

[a]Minimum edible Kcal/day per capita

landscape. Shape–function relations are very important,[4] as we can see considering functional configurations.

The *functional* subsystems with *specific configurations within the ecotissue* formed by ecocoenotopes are named *Landscape Apparatuses LA* (Ingegnoli 2002, 2015). The most important are: HGL Hydro-Geologic, RNT Resistant (elements with high metastability, e.g. forests), RSL Resilient (elements with high recover capacity, e.g. prairies or shrub lands), PRT Protective (elements which protect and compensate other elements or parts of the mosaic), PRD Productive (elements with high production of biomass, e.g. agriculture), SBS Subsidiary (systems of human energetic, transport and work resources), RSD Residential (systems of human residence and its dependent functions).

Strictly related is the *vital space* per capita [m²/ab] (Ingegnoli 2015), intended as the set of portions of the landscape apparatuses within the examined LU indispensable for an organism to survive, better known as *Standard Habitat* per capita (SH).

It is available for an organism (man or animal), divisible in all its components, biological and relational. A *minimum theoretical standard habitat* per capita (SH*), both for human and animal population, have been estimated.[5]

Note that, even for the same species, SH may change in function of the bioclimatic belt and the landscape type.[6] In Table 1, the SH* in relationship to human population and the main climatic belts of the biosphere are exposed, even if values can be locally updated.

[4]See also: E. Del Giudice, A. Tedeschi, *Lo sviluppo spontaneo della conoscenza negli organismi viventi. Unità di funzione e struttura,* "Rivista di Filosofia Neo-Scolastica", CVI, pp. 537–544.

[5]In function of the minimum edible Kcal/day per capita [1/2 (male + female diet)]; the productive capacity (PRD) of the minimum field available to satisfy this energy for 1 year, taking into account the production of major agricultural crops; an appropriate safety factor for current disturbances; the need for natural and/or semi-natural protective vegetation for the cultivated patches (Ingegnoli 2015: 61–64).

[6]In the case of human populations (idem), we will have a SH_{HH}, that is a SH referred to the human habitat (HH): $SH_{HH} = (HGL + PRD + RES + SBS + PRT)/N°$ of peoples [m²/inhabitant].

3.2 Landscape Processes (Physiology)

Difficult to be synthesized, a focus on four aspects is presented:

1. Each living entity from the local to the upper scales manage *a flux of energy to reach and maintain a proper level of organization and structure, through its vegetation communities*[7]: a landscape systemic function, named *Bionomics Territorial Capacity of Vegetation* (BTC) (Ingegnoli 2002, 2005, 2011, 2015), linked to metastability (based on the concept of resistance stability) gives us *a quantitative evaluation of this flux of energy,* through two coefficients to measures the degree of the relative metabolic capacity a_i and the degree of the relative antithermic (i.e. order) maintenance *bi* of the same main vegetation communities.

   Note that the BTC is a crucial systemic function: ranges (standard deviations) of BTC values can be measured, following the LaBiSV methodology related to both natural and human eco-bionomics systems. These data can be useful in LU analysis and landscape assessment. An example is shown in Fig. 8.
2. *Humans affect and limit the self-regulation capability* of natural systems. An evaluation (systemic % of LU surface) of this *ability* brings to the concept of *Human Habitat (HH).* Ecologically speaking, the HH cannot be the entire territorial (geographical) surface: it is limited to the subsystem of human ecocoenotopes in landscape units (e.g. urban, industrial and rural areas) and to the semi-human ones (e.g. semi-agricultural, plantations, ponds, managed woods).

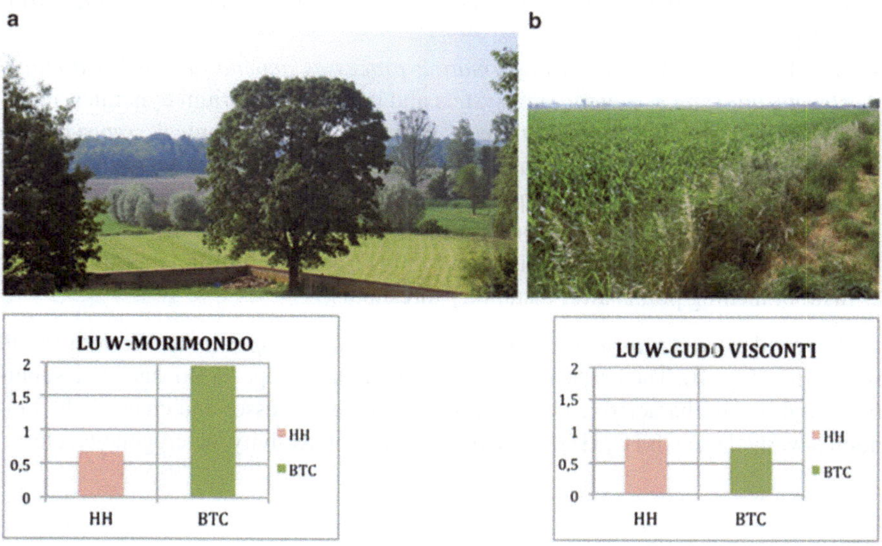

Fig. 8 Comparison between two agrarian landscapes near Milan: note the sharp difference in the BTC values, related to the different organization levels of these landscape units (left, well organized corresponding to high BTC value; right, lower level of BTC for an unstructured one)

[7]Their metabolic data (biomass, gross primary production, respiration, B, R/GP, R/B) in Ingegnoli, 2002: 113.

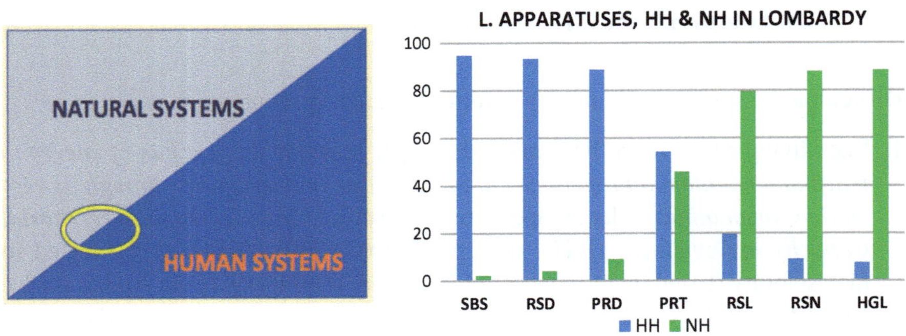

Fig. 9 The comparison between the reductionist concept of natural vs. human components in ecology (left) and the systemic concept (right). The difference is sharp

The *Natural Habitat (NH)* is the subsystem of natural ecocoenotopes, with dominance of natural components and biological processes, capable of normal self-regulation. In Fig. 9 the sharp difference between the reductionist concept of natural and human components vs. the systemic one, in which the landscape apparatus and the human (HH) and natural (NH) habitats are integrated, is underlined: they are expressed by two non-linear functions. The absence of HH or NH can be possible only as an exception, e.g. in a wide industrial area.

3. Consistently with Systemic Theory, a landscape unit (LU) can be *autotrophic or heterotrophic* on the basis of its *Systemic Carrying Capacity (σ)* related to the *Ratio SH/SH** [where SH* is the theoretical minimum] (Ingegnoli 2002, 2011, 2015).
4. In a landscape the main *transformation processes depend on the hierarchical structuring of an eco-bionomics system* and its non-equilibrium dynamics, metastability, coevolution, evolutionary changes and ecological *reproduction* (Ingegnoli 2002, 2015). The basilar HH needs imply a *transformation impact* on natural systems, which consequently suffer a *Transformation Deficit (TD)* that must be compensated with opportune protective (PRT) systems. *Succession* does not work as linear and mechanistic, but follows non-equilibrium thermodynamic with branching points after instability thresholds.

A typical *altered transformation of agricultural landscapes* in the Western World is shown in Fig. 10. The passage from a traditional landscape "a Bocage" to a suburban–rural one is characterized by destructive human processes and changes the bionomics variables, as BTC, HS/HS*SH/SH*, HH, connectivity, heterogeneity, etc.

3.3 Landscape Health State Diagnosis

Some considerations on landscape pathology. The definition of landscape as a specific level of life organization becomes a challenge for environmental evaluation, first of all because man has to pass from a discipline related to technology, economy,

Pathogenic scheme of the agrarian industrialization syndrome of temperate agricultural landscape of the plains.		
Original State Permanence: 3 to 18 centur	**Traditional agricultural landscape "a Bocage"** (BTC = 1.2-2.1 Mcal/m^2/yr; HS/HS*= 2.5-6.0; HH = 50-75%) heterogeneity, connectivity and circuitry = good	
Main human cause of alteration	Socio-economic pressure due to growing agrarian production	Increase of help for agrarian technologies
	Specialization of cultivations	Canalization of small rivers
	Increase of arable land and of chemical fertilizers	Cutting of tree lines and hedgerows, mechanical irrigation
Positive feedback processes	Destruction of forest patches, Increase of crop pests, Increase of chemical pesticides, Decrease of fauna, Soil depletion	
	Enlarging field area, increase road network	Rupture of geomorphologic constraints,
Altered State Permanence: 20 to 80 yr	**Open mono-cultural landscape** (BTC = 0.9-1.3 Mcal/m^2/yr; HS/HS*= 4-9; HH = 70-85%) heterogeneity, connectivity and circuitry = weak, partial	
	Structural weakness, Increase system fragility, Attraction for highways, Attraction for industrial areas, Increase of fragmentation	
Disordered State Permanence ?	**Suburban-rural landscape** (BTC = 0.7-1.1 Mcal/m^2/yr; HS/HS*= 0.8-2.7; HH = 80-90%) heterogeneity= increasing; connectivity and circuitry = disrupted	
	Loose of functionality, Decrease of agrarian production	

Fig. 10 The passage from a traditional landscape "a Bocage" to a suburban–rural one. Note the possibility to quantify the changes in systemic parameters and to refer to specific landscape syndromes within a clinical screening

sociology, urban design, visual perception *and* ecology, to another related to bionomics, natural sciences, medicine *and* traditional disciplines. This change implies the use of medical terms, first of all the concept of health, to be adopted also for a landscape unit. The most known definition of health is due to WHO (World Health Organization): "A resource for everyday life, not the objective of living. Health is a positive concept emphasizing social and personal resources, as well as physical capacities" with a recent add "the ability of a body to adapt to new threats and infirmities". This is not a complete definition, but it has significance that medicine should be referred to the concept of health not only to disease. The state of health regards also a condition of *normality*, indispensable to recognize a state of alteration, at all levels of the biological organization.

More problematic may be the extension of the concept of pathology, because its definition in dictionaries is only partial, e.g. "the study of the way diseases and illnesses develop" (Collins) or "the science of the causes and effects of diseases, especially the branch of medicine that deals with the laboratory examination of samples of body tissue for diagnostic or forensic purposes" (Oxford). As underlined by Konrad Lorenz (1978), the difficulty to understand the concept of pathology lays in its inextricable link with physiology, within which to understand a function we need its alteration and vice versa, with continuous feedbacks. Thus pathology is an alteration of the physiology and behaviour of a living entity able to reduce and modify negatively its normal functionality and to induce physiological reactions: pathology may end with recovery or death or adaptation to a different physiology.

Despite the valid intuitions of Von Humboldt (1846), on the concept of Physiology of a Territory, the pervasive reductionism of general ecology didn't and doesn't consider it, notwithstanding Landscape Bionomics arises. In this case the discipline of landscape bionomics is indispensable, having the capacity to analyse both processes related to a landscape unit or its lower levels.

The study of the pathology of any living system, independently from the levels of scale and organization, needs a basic clinical-diagnostic methodology, which cannot be avoided. This is true also for landscape dysfunctions and it may be articulated in six phases: survey of the symptoms; identification of the principal causes; analysis of the reactions to pathogen stimuli; risks of ulterior worsening; choice of therapeutic directions; control of the interventions.

Like in medicine, environmental evaluation needs comparisons with "normal" patterns of behaviour of a *system* of ecocoenotopes. Therefore, the main problem becomes how to know this normal state and/or, at the same time, the levels of alteration of that system.

After the study of 45 landscape units (mainly in North Italy), an exceptional correlation between the Biological Territorial Capacity of Vegetation and the Human Habitat—that is between *the flux of energy needed by a living system to reach and maintain a proper level of organization and structure* (BTC) and *the measure of the humans control and limitation* (HH) *of the self-regulation capability* of natural systems—was found, with an $R^2 = 0.95$ and a Pearson's correlation coefficient of 0.91[8]: so, it was possible to build the simplest mathematical model of bionomic normality (Fig. 11), available for *a first framing of the dysfunctions* of landscape units.

Below normal values of *bionomic functionality* (BF = 1.15–0.85), with a tolerance interval (0.10–0.15 from the curve of normality) we can register three levels of altered BF: altered (BF = 0.85–0.65), dysfunctional (BF = 0.65–0.45) and degraded (BF < 0.45). The vertical bars divide the main types of landscapes, from Forest-Natural (high BTC natural) to Dense-Urban: each of them may present a syndrome.[9] Again, this model is indispensable to reach a first eco-bionomics diagnosis on the health of an examined landscape unit (LU), to control the effects of a territorial planning design, to study the landscape transformations, etc.: note that it is a complex one, because both HH and BTC aren't two simple attributes and their behaviour is not linear.

This methodology cannot evaluate the modifications on specific elements of the system, it can't be mechanistic. A systemic diagnostic method has to be "spectral", as underlined by Giuliani (2015). The physician will examine the patient through a neutral spectrum of experimental probes: the diagnosis will emerge as a correlation structure induced by the peculiar syndrome among different probes. Each probe (symptom) has in general (no or) low value per se, while the entire profile or "diagnostic frame" acquires a clinic significance just because it induces a correlation

[8] (About three times the minimum value of significance.)

[9] For the articulation of landscape pathologies, see (Ingegnoli 2015, Sect. 4.5:100–110).

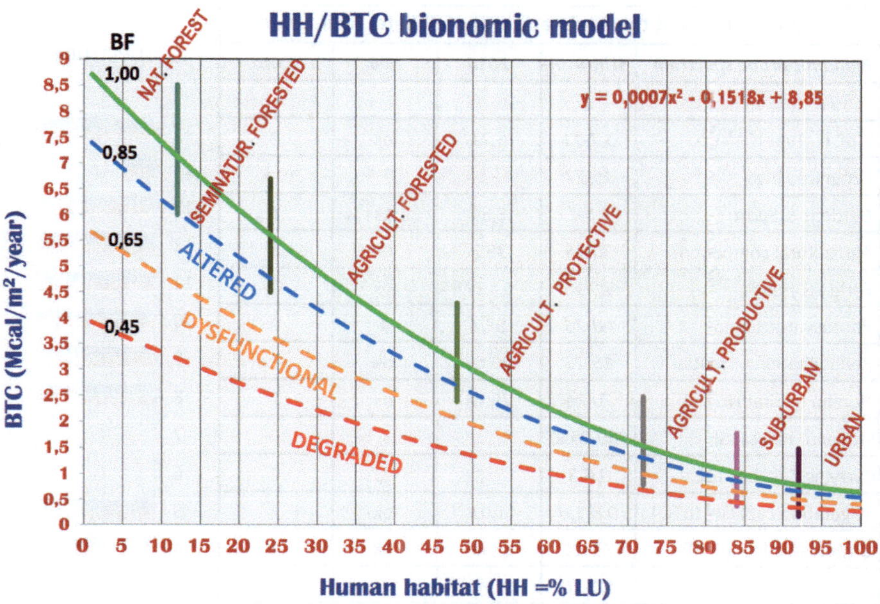

Fig. 11 The HH/BTC model, able to measure the bionomics state of a LU. Dotted lines express the BF level, that is the bionomics functionality of the surveyed LU

structure among different parameters. Remember that "syndrome" means to run together and that the best scale of explanation is the one capable to maximize the importance of the relations among the elements, both natural and human, thus to permit a valid diagnosis and control of an upper living system, confirming the role of the eco-bionomic scientist as "*ecoiatra*" (Ingegnoli 2011, 2015): thus the crucial importance of the landscape even with regard to Human Health.[10]

A deepen method of Diagnostic Index (DI) can be elaborated using 12–18 eco-bionomics variables and parameters, as exposed in the mentioned books (Ingegnoli 2011, 2015). In many case studies the result may find a more critical BF, about 10–15% worst than in this HH/BTC model. After having established the type of landscape in examination, one must quantify the "*distance*" of each ecological variable from the threshold of normality, giving a score, according to the offset values (%) (Fig. 12). This means *using a diagnostic index* (DI).

The diagnostic index (DI) is given by the total score divided by the number of estimated parameters multiplied by the maximum score of 2. Evaluation Scores: 0–10 = 2; 10–30 = 1; 30–60 = 0.5; >60 = 0. The reference intervals for the expression of the scores are exposed in Table 2.

[10]The landscape dysfunctions are correlated with the increase of mortality rate MR, independently from pollution. All the environmental alterations are registered as 'stressors' by a basilar ethological alarm process. So, bionomic landscape dysfunctions may attempt our health reducing our body defences. The risk factor in premature death can be elevated (Ingegnoli 2015: 110–115).

Diagnostic Index of the landscape unit of Bollate, near Milan				
Bionomic probe spectrum	*SUR-RUR*	*2010*	dev	score
POP/1000 (inhabitants)		**35,75**		
LTpE (L. Type)	*5,7-11*	5,71	ok	**2**
Urbanized area	*15-27*	39,84	47,4	**0,5**
Gardens & Sport	*3,5-7*	8,84	ok	**2**
Agricultural components	*30-55*	38,2	ok	**2**
Forest & Heath	*8,0-25*	12,27	ok	**2**
Photosyntetic area	*60-70*	57,2	4	**2**
Rehabilitation potential	*15-25*	20,66	ok	**2**
Human habitat HH (%)	*70-86*	79,84	ok	**2**
Standard habitat SH	*800-1650*	293	63,4	**0**
Carrying Capacity (σ)	*0,6-3*	0,2	65	**0**
Vegetation capacity (BTC)	*0,8-1,6*	1,03	ok	**2**
L. Metastability (g-LM)	*7,5-15*	7,52	ok	**2**
territorial CBSt	*8,0-20*	5,8	27,5	**1**
forest & shrub CBSt	*22,5-30,5*	14,86	34	**0,5**
Potential core areas	*20-35*	21,4	ok	**2**
total score	*27,2-32*			22
Diagnostic Index (DI)	*85-100*			**73,33**
Bionomic state evaluation	*normal*			**altered**

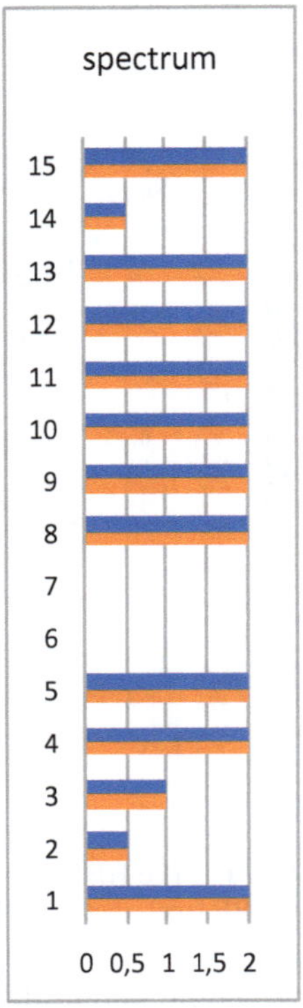

Fig. 12 "Diagnostic frame" of a landscape unit near Milan. Each probe has in general low value per se, while the entire profile acquires a clinic significance

Table 2 Diagnostic evaluation of a landscape after the measure of the D.I. from Ingegnoli (1991)

Pathology levels	Diagnostic index (DI)	Diagnostic evaluation	Physiological-pathological notes	Ecological health and interventions
I	0.85–1.00	Normal	Homeostatic plateau	Quite good health, only prevention
II	0.65–0.85	Alteration	Compensation needed	Instable health, some therapies and intervention
III	0.40–0.65	Dysfunction	Some physiological damages	Dysfunction, wide intervention needed
IV	0.10–0.40	Severe dysfunction	Harmful effects	High dysfunctions, difficult intervention
V	<0.10	Extinction	Irreversible damages	Degenerative transformations

As former expressed, the entire profile of probe spectrum or "diagnostic frame" acquires a clinic significance just because it induces a correlation structure among different parameters. So, DI is the first significant and important step of diagnostic screening, but the process needs deeper considerations and a good clinical eye: a diagnosis is not a deterministic process! We should remember that it's the relationship between "pathology" and "physiology" of the system that allows a diagnosis in the clinical sense of the landscape in question: one must understand how the system moves from the state of normality due to pathogenic stimuli and, with a projection of the information, to assess where the damage could rise to the structure and functions at a time congruent.

3.4 *Planetary Health Control*

The main impacts of anthropogenic change on human health are generally exposed by Planetary Health Alliance through a process in five steps: (A) underlying drivers, (B) ecological drivers, (C) proximate causes, (D) mediating factors, (E) health effects. This process follows conventional ecology (see Sect. 2.3), so for each one of the five steps the next aspects are put in evidence: (A) Consumption, Demographic shifts, Technology; (B) Global pollution, Climate change, Resource scarcity, Land use/cover change, Altered biogeochemical cycles, Biodiversity loss; (C) Air quality, Food production, Infectious disease exposures, Access to freshwater, Natural hazards; (D) Governance, Wealth, Philanthropy, Technology, Culture/Behaviour; (E) Malnutrition, Infectious disease, Non-communicable disease, Mental health, Displacement and conflicts.

These components are no doubt appropriate, but the lack of the bionomics elements (and principles) is evident, even after the synthetic exposition in this work. In facts we must add:

(a) To underlying drivers, *Neoclassical Economy*, today dominating our Globe, primordially indebted to nature (the exchange values derived from labour cannot be the economic measure of everything!).
(b) To ecological drivers, *Living complex systems* (e.g. landscape units), the hierarchical organization of Life on Earth (well farther biomes, populations and communities), that is the unitary frame within which all the previous mentioned aspects can be integrated.
(c) To proximate causes, *Landscape bionomics degradations and artificialization*, which implies the Diagnosis of Landscape Unit Syndromes and Artificialization.
(d) To mediating factors, *Religions.*
(e) To health effects, *Premature death by systemic diseases*, growth of morbidity due to environmental Stress (Ingegnoli and Giglio, 2017) and *feedback* to artificialization.

Note that the most important addiction (Fig. 13) is the point (b), from which four groups of alterations derive, each of one leading to human health alterations: (1)

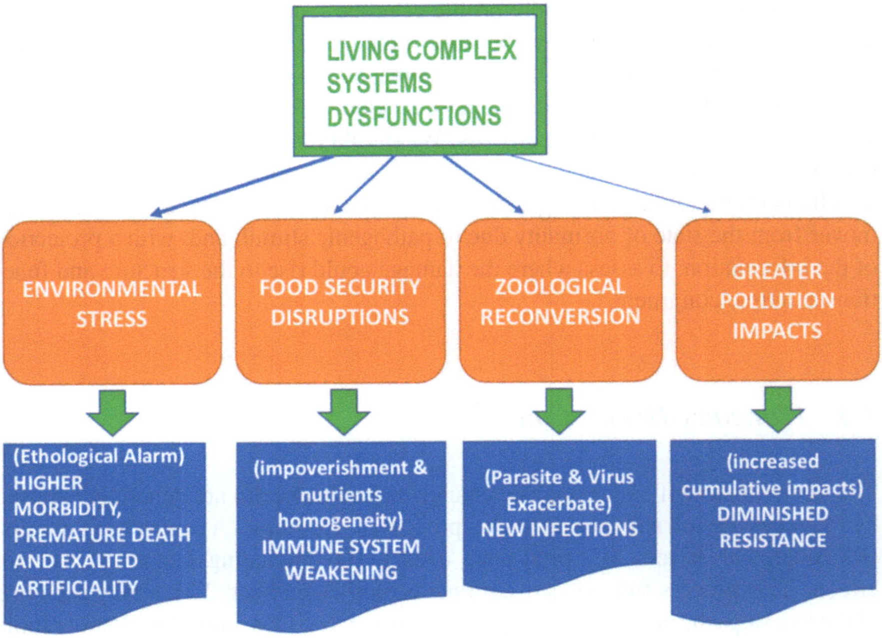

Fig. 13 Main consequences derived from living complex systems dysfunctions, the most important systemic component of ecological drivers

Environmental stress (ethological alarm) ≫ higher morbidity, premature death and exalted artificiality; (2) Food security disruptions (impoverishment and nutrients homogeneity) ≫ immune system weakening; (3) Zoological reconversion (parasite and virus exacerbating) ≫ new dangerous infections; (4) Greater pollution impacts (increased cumulative impacts) ≫ diminished resistance. Moreover, the mentioned process A–E has not to be presented as a linear sequence of causal determination, but as a network of complex systemic graph.

4 Conclusion

The title of this chapter is clear: "Is Present Ecology a Systemic Discipline? New Scientific Paradigms Lead to Bionomics". If the basilar concept of Ecology is the "Ecosystem" is this a desecrating title? The answer depends on considerations related to the challenges come up by Reality, which is complex, creative and needing freedom. Many epistemological studies indicate that the modern scientific method, i.e. Galilean, deductive, experimental, although not yet been passed, has shown severe limitations. First of all, it is not able to guarantee the certainty, because it remains substantially imperfect, as demonstrated by Gödel's theorems of incompleteness and Tarski's one of indefinability. In fact, all scientific disciplines have found themselves in crisis, more or less deep, in the course of the past century, which has even shown the flourishing of an exceptional development in all branches of science.

These facts drove Science towards a change of its fundamental principles, from a too dominating reductionism to a complex systemic basis, therefore to new scientific paradigms. That's the reason of the arise of new biological disciplines, more available to study complex systems, e.g. epigenetics, agroecology, systemic medicine, bionomics, and environmental health. All of them underline the limits of ecology in studying complex systems: we have shown the *ambiguity* of the concept of *ecosystem*, a reinterpretation of *biodiversity* and *resilience*, the non-systemic concept of vegetation *association* and the mistakes in the delimitation of a *metropolitan* area. So, what's Bionomics? It's the new discipline investigating the *Laws of Life on Earth* as a hierarchical organization of complex systems, acting as living entities: it transforms many principles of conventional ecology and confirms the preeminent importance of the systemic approach to correctly evaluate and care the Planetary Health, giving it a wide theoretical corpus. The brief synthesis of some of the main aspects of Bionomics and Landscape Bionomics were given, especially underlining the diagnosis of a landscape unit.

In conclusion, we must confirm and underline that each living entity must be investigated in its physiology (e.g. metastability, biologic functions, autotrophy etc.) and pathology by the discipline of Bionomics and Landscape Bionomics. This is to check "if", "how" and "how much" living system alterations could reflect on human health, even independently from pollution.

References

Agazzi, E. (Ed.). (2014). *Science, metaphysics, religion*. Milano: F. Angeli. (Chap. I).

Bailey, R. G. (1996). *Ecosystem geography*. New York: Springer.

Besedovsky, H. O., del Rey, A., & Sorkin, E. (1981). Lymphokine-containing supernatants from con A-stimulated cells increase corticosterone blood levels. *Journal of Immunology, 126*, 385–387.

Braun Blanquet, J. (1926). *Etudes phytosociologiques en Auvergne*. Clermont Ferrand: Mont-Louis G.

Bronowski, J. (1969). *Nature and knowledge: The philosophy of contemporary science*. Condon lectures, London, Oregon State System of Higher Education.

Crick, F. (1970). Central dogma of molecular biology. *Nature, 227*, 561–563. Baltimore, D. (2001). Our genome unveiled. *Nature, 409*, 816.

Einstein, A. (1944). *Einstein archives* (pp. 61–574).

Einstein, A. (1949). *Autobiografisches,* (© 1949, 1970) Library of Living Philosophers, Chicago. *tr.it. Autobiografia scientifica*, in. A. Einstein Opere scelte, Bollati- Boringhieri, Torino.

Giglio, E. (2002). Synthetic glossary. In V. Ingegnoli (Ed.), *Landscape ecology: A widening foundation* (pp. 323–333). Berlin: Springer.

Giuliani, A. (2015). Complessità Organizzata. In L. Urbani Ulivi (Ed.), *Strutture di Mondo, il pensiero sistemico come specchio di una realtà complessa* (Vol. III). Il Mulino.

Ingegnoli, V. (1971). Ecologia territoriale e progettazione: significati e metodologia. In *L'ingegnere di fronte alla sopravvivenza umana, Milano, Coll. Ingegneri* (Vol. I, pp. 398–400).

Ingegnoli, V. (1980). *Ecologia e progettazione* (p. 310). Milano: CUSL.

Ingegnoli, V. (1991). Human influences in landscape change: Thresholds of metastability. In O. Ravera (Ed.), *Terrestrial and aquatic ecosystems: Perturbation and recovery* (pp. 303–309). Chichester: Ellis Horwood.

Ingegnoli, V. (2001). Landscape ecology. In D. Baltimore, R. Dulbecco, F. Jacob, & R. Levi-Montalcini (Eds.), *Frontiers of life* (Vol. IV, pp. 489–508). New York: Academic.

Ingegnoli, V. (2002). *Landscape ecology: A widening foundation* (pp. XXIII + 357). Berlin, New York: Springer.

Ingegnoli, V. (2005). An innovative contribution of landscape ecology to vegetation science. *Israel Journal of Plant Sciences, 53*, 155–166.

Ingegnoli, V. (2011). *Bionomia del paesaggio. L'ecologia del paesaggio biologico-integrata per la formazione di un "medico" dei sistemi ecologici* (pp. XX + 340). Milan: Springer.

Ingegnoli, V. (2015). *Landscape bionomics. Biological-integrated landscape ecology* (pp. XXIV + 431). Heidelberg, Milan, New York: Springer.

Ingegnoli, V., & Giglio, E. (2005). *Ecologia del Paesaggio: manuale per conservare, gestire e pianificare l'ambiente* (pp. 685 + XVI). Napoli: Sistemi editoriali SE.

Ingegnoli, V., & Giglio, E. (2017). Complex environmental alterations damages human body defence system: A new bio-systemic way of investigation. *WSEAS Transactions on Environment and Development, 13*, 170–180.

Ingegnoli, V., & Pignatti, S. (2007). *The impact of the widened Landscape Ecology on Vegetation Science: Towards the new paradigm* (s. IX, Vol. XVIII, pp. 89–122). Springer Link: Rendiconti Lincei Scienze Fisiche e Naturali.

Lorenz, K. (1978). *Vergleichende Verhaltensforschung: Grundlagen der Ethologie*. Berlin, Wien: Springer.

Lovelock, J. (2007). *The revenge of Gaia: Earth's climate crisis & the fate of humanity*. London, New York: Penguin Books.

Lovelock, J., & Margulis, L. (1974). *Atmospheric homeostasis by and for the biosphere: The Gaia hypothesis*. Tellus XXVI.

Naveh, Z., & Lieberman, A. (1984). *Landscape ecology: Theory and application* (pp. XXVII + 360). New York: Springer.

O'Neill, R. V., De Angelis, D. L., Waide, J. B., & Allen, T. F. H. (1986). *A hierarchical concept of ecosystems*. Princeton: Princeton University Press.

Odum, E. P. (1971). *Fundamentals of ecology* (3rd ed.). Philadelphia: WB Saunders.

Odum, E. P. (1983). *Basic ecology*. CBS College Publ. USA.

Pignatti, S. (1995). *Ecologia vegetale*. Torino: UTET.

Pignatti, S., Box, E. O., & Fujiwara, K. (2002). A new paradigm for the XXIth Century. *Annali di Botanica, II*, 31–58.

Rivas-Martinez, S. (1987). Nociones sobre Fitosociologıa, biogeografıa y bio-climatologia. In *La vegetacion de España* (pp. 19–45). Madrid: Universidad de Alcalá de Henares.

Tansley, A. G. (1934). Editorial note, to Michelmore's Letter to the Editor. *Journal of Ecology, 22*, 317.

Tüxen, R. (1956). Die heutige potentielle naturliche Vegetation als Gegenstand der Vegetationkartierung. *Angew. Pflanzensoziologie Stolzenau/Weser, 13*, 5–42.

Von Humboldt, A. (1846). *Kosmos*. Stuttgard.

Waddington, C. H. (1942). The epigenotype. *Endeavor, 1*, 18–20 (Reprinted in *Int J Epidemiol* 2012, 41, 10–13).

Waddington, C. H. (1957). *The strategy of the genes: A discussion of some aspects of theoretical biology* (pp. ix + 262 pp). Allen & Unwin ed., London.

Zonneveld, I. S. (1995). *Land ecology*. SPB Amsterdam: Academic.

Architecture and Systemics: A Brief Outline

Carlotta Fontana

Abstract Architecture is a complex subject in itself, as it shapes the built environment where people live, answering to human needs, expressing the manifold levels of values which define society in its culture, economy, and politics. During the twentieth century, a number of design theories, both in Europe and in the USA, linked Architecture and Complexity drawing inspiration from Systemics, Information theories, and Cybernetics. Thus, being closely connected to industrial production, the main goal was to reduce uncertainty in the design process, promoting optimization. In the industrial design process, a sequence of requirements defines the exact level of fitness-for-purpose of a product. Such ideas proved to be unsuitable for many architectural design purposes: "functional optimization" can be applied to an object, a device, and a machine; it seems to be useless, and even dangerous, when applied to an evolutionary entity as the built environment seems to be.

This chapter endeavours to trace an outline of this difficult relationship.

1 Complexity Made Simple?

The very nature of Architecture is complex. The understanding of such complexity has accompanied the development of theory since the Vitruvian Triad. To name just one masterpiece of twentieth century architectural criticism, the integrated theory of architecture expressed by Christian Norberg-Schulz's *Intentions in Architecture* (Norberg-Schulz 1965) represents a truly systemic comprehension of architecture without ever naming the word.

More specific references to Systemics and systemic thought and language came in when the industrialization of building process approached its maturity, after WW2, even if some clues could be traced back to the industrial revolution, when a

C. Fontana (✉)
Dipartimento di Architettura e Studi Urbani—DAStU, Politecnico di Milano, Milano, Italy
e-mail: carlotta.fontana@polimi.it

L. Urbani Ulivi (ed.), *The Systemic Turn in Human and Natural Sciences*,
Contemporary Systems Thinking, https://doi.org/10.1007/978-3-030-00725-6_5

number of completely new problems faced the architectural profession and challenged the theory.

For centuries before industrialization, Architecture had represented a high-rank applied art, quite often supported by highly refined formal prescriptions, always supported in the construction phase by robust technical knowledge improved by experience during time. The practice of Architecture was destined to major buildings, promoted by public interest or wealthy clients, representing multiple social, civic, and religious values. The main part of the ordinary built environment (Rudofsky 1964, 1977; Habraken 1998) grew up in layers over time—without neither architects nor engineers—taking shape according to the geo-climatic peculiarities of the place and to the local resources, activities, customs, and technical skills. At all scales, the construction process was slow, the means and materials mostly local, the knowledge and techniques improved over time by trial and error and handed down by tradition, through apprenticeship.

Industrialization, speeding up all human activities, undermined the foundation of architectural culture and knowledge. At the end of the 19th Century, massive urbanization forced architectural culture to face unusual problems, under different aspects. An important and much-debated question was how to express the new aesthetic and symbolic values of the industrial age. The most unusual problem was how to design large quantities of low-cost housing of acceptable quality, new services, and urban equipment to meet the collective needs of a changed society. Most architectural culture, after World War 1, was committed to defining the minimum housing requirements to accommodate masses of new clients, both numerous and unknown.

The studies by Alexander Klein in Berlin, those by Grete Schütte-Lihotzky in Frankfurt in the 1920s, to name just a few, tried to integrate Taylorist-inspired ideas into the design process, with the objective of giving everyone an efficient, comfortable, and pleasant home despite the financial constraints. These designers analysed the usual activities that take place in the house, measuring time and ergonomic relationships between movements, paths, and equipment, committed to the idea of improving the efficiency, the health, and the well-being of their unknown and anonymous "clients". This meant applying the industrial conception of functional analysis and organization to the production and reproduction of labour power to the activity that customarily take place in the environment where a family lives.[1]

In order to satisfy the housing needs of this new *mass-entity*, it was not possible to investigate the needs of a specific client. It became essential to trace—or to imagine—the significant elements common to countless, faceless individuals whose customs and ways would be increasingly levelled out by life in the industrial city. These people were identified as "users", expected to find satisfaction by living in well-equipped functional spaces. The study of repeatable typological solutions, suitable for buildings constructed by means of fast techniques and new materials available through industrial production, implied the "construction" of an average user, whose

[1] Studies on the *Existenzminimum*, as it was termed in German, were carried out in the 1920s both in capitalist Europe and in the newborn Soviet Union, with different degrees of insight about the women's role.

uniform behaviour and aspirations represented the foundation of the industrial and rationalist idea of *standard*.

"All men have the same organism and the same function. All men have the same needs", claimed Le Corbusier (1923) while, as early as in 1932, Hitchcock and Johnson (1932) criticized the idealism of the European functionalists, remarking that they aimed to satisfy the needs that one should have, rather than actual ones: "Functionalism is absolute as an idea rather than a reality (…) The Siedlungen implies preparation not for a given family but for a typical family. This statistical monster (…) has no personal existence and cannot defend himself against the sociological theories of the architects (…) Europeans build for some proletarian superman of the future".

Sigfried Giedion's *Mechanization Takes Command* (Giedion 1948) focused the question of mass-production of buildings and bore a significant sub-title: "a contribution to anonymous history". Giedion examined the effects of mechanization in everyday life, tracing the outline of a social history of technology, which at the time represented a critical breakthrough. He invoked the creation of "chairs of anonymous history" in the University and blamed as "murder of history" the destruction of documents about the early stages of industrialization, claiming that inventions, mass-production, and the work of ordinary people in the industrial era "are continually shaping and reshaping the patterns of life" in an unprecedented way, at every possible level. Giedion also suggested to open a research field to find an answer to the question: "what does mechanization mean to man?", and to investigate such topics as the dangers of losing human control over products and of increasing dependence upon industrial production, in a situation where, in general, "man is overpowered by means".

Marking a significant distance from his previous work (Giedion 1941), *Mechanization Takes Command*, published shortly after the apocalypse of WW2, suggested the analogy between mass-production and mass-destruction, and recalled the horrors of organizational efficiency applied to extermination. Thus, while claiming a well-balanced attitude towards the historical condition of "mechanization", Giedion questioned the optimistic, positive aura surrounding the idea of progress itself: after WW2, "men have become frightened by progress, changed from a hope to a menace (…) before our eyes our cities have swollen into amorphous agglomerations. Their traffic has become chaotic, and so has production". Giedion would not reject the notion of mechanization; he rather aimed to defining mechanization's place in history, society, and in culture, while rejecting the mechanistic conception of the world. Such conception, he argued, had been swept off every cultural domain already—from physics to biology, psychology, and art. He rather suggested a systemic, *holistic* way of conceptualizing "domains having to do with the human organism" and closed his book with a list of "new balances" required: balance between individual and community, between the world as a whole entity, and local issues, between the spheres of knowledge, and "between the human organism (…), its organic environment and its artificial surroundings".

Heavy traces of a "mechanistic systemic thought" show in the 1950s and 1960s rational design theories developed in the USA and the UK, drawing on the experience gained in industry to reduce errors, uncertainty, risks, costs, and time.[2]

During the 1940s and in war production, a number of techniques of analysis and control for various processes—planning, industrial design, and production—had been developed. Reduction of error entails the capability to integrate and manage the relationships and information flows between different actors in a complex process. Decision-making techniques were deployed along the lines of Operational Research (OR), which represents a method of mathematical analysis to identify and break down one specific general problem in sub-problems, in order to define a sequence of decisions capable of achieving *performance improvement* in both the process and its final product. Thus defined, the decision sequence can be summarized in a mathematical model that allows evaluating different solutions by modifying certain variables (Broadbent 1973).

The rational methodology, as applied to the programme/project/production flow, refers to information theories and cybernetics[3], the science of control and communicationin animals and machines (Wiener 1948; Ashby 1956) and focuses its analysis on the relationships between the elements of a system and their role.

During the 1960s, the ideas of *input, output,* and *feedback* became familiar to rational architectural design, with different regional variations between European countries and the USA. Morris Asimow (1962) outlined a method describing industrial design in terms of information process, whose steps subsequently gather, handle, and organize information in a creative way. Such process has an iterative character and prescribes the derivation of decisions which must be optimized, communicated, and *"tested or otherwise evaluated"*.

Accordingly, rational design processes were generally structured in phases modelled on a decision sequence with feedbacks, often represented by flow diagrams.[4]

Complexity more directly approached the world of architecture via the Hochschule für Gestaltung established in Ulm in 1949.[5] The Ulm School promoted a system-based, formalized approach to architectural design, combining the Bauhaus commitment to artistic production for the industrial age and the optimization aims of Operational Research. Along this line, the relationship between humans

[2] Main studies in the Anglo-American area were: M. Asimow, *Introduction to Design*, 1962; J.C. Jones, "A Method of Systematic Design", 1963, in: *Design Methods*, 1970; S.A. Gregory, *The Design Method*, 1964; L.B. Archer, *Systematic Method for Designers*, 1965.

[3] Cybernetics, recalling the assertions of contemporary science on the impossibility of studying complex systems by reducing them to their simplest components, searches for methods capable of analysing and controlling systems of *extreme intrinsic complexity*.

[4] Broadbent (1973, p. 257).

[5] Tomàs Maldonado, professor at the Ulm school from 1954 to 1967 directed it from 1956 to 1960, establishing the disciplinary and academic field of *Environmental Design*, within the frame of a wider "design philosophy" based on analytical methodologies. He had a fundamental influence on design theories in Italy; he was professor of Environmental Design at the University of Bologna (1976–1984) and at the Politecnico di Milano (1985–1994) where he greatly contributed to establish the school of Industrial Design.

and human-designed-and-built environment at different scales entailed scientific analysis and a design approach where a number of variables link the corresponding users' needs—which are closely connected to their environment—and the functional requirements of their activities.[6]

The approach called *metadesign*[7] represented a formalization of the design process which could generate models of design behaviour apt to deal with uncertain and changing situations. It was conceived as an "ordered set of operations to achieve congruence between premises and conclusions, through systematic processing tools, and to knowingly define the limits of design alternatives compatible with the problem" (Boaga and Giuffrè 1975). The procedure takes into account both the analytical phase and the synthetic, conceptual one, providing "the organization of a system of spatial requirements descending from human activities, both specific and in their mutual relationship, which by concretizing and quantifying these requirements in relation to any specific context, brings forward a field of design variations (dimensional, typological, etc.) from which solutions can be derived that correspond to the general objectives of the customer and user" (Magnaghi 1973).

Generally speaking, the rational design approach proceeds from the preliminary analysis of the users' needs according to the system of activities to be provided, to the definition of a programme containing specific requirements, to the construction of a model representing an environmental spatial system which properly meets the requirements of the organization of the established activities. The environmental subsystem of spaces, and the technological subsystem physically containing it, represent a complex *building organism*: a dynamic system that, in performing its functions, continually processes matter, energy, and information that flow in and out of its physical boundaries.

At this stage, rational design theories[8] agreed that the designer's goals should be expressed in terms of *performance* which had to be specified in a set of criteria. The conjoined terms of *need-requirement-performance* were at the core of this idea.

Under a different point of view, the need of a more formalized method to help design accomplish the new tasks posed by mass-building production represented an updated version of the old debate about Architecture being disputed by the realms of Art and Science. J.C. Jones wrote: "The method is primarily a means of resolving a conflict that exists between logical analysis and creative thought. The difficulty is that imagination does not work well unless it is free to alternate between all aspects of the problem, in any order and at any time, whereas logical analysis breaks down if there is the least departure from a systematic step-by-step sequence (…) so systematic design is primarily a means of keeping logic and imagination separated by external rather than internal means".[9] Jones's assumptions were widely shared, in a

[6] In Italy, this approach to design in architecture gave life to the academic discipline "Tecnologia dell'Architettura" (Architectural Technology), established in 1969.

[7] Andreis Van Onck brought forward the idea while at ULM in 1963.

[8] Broadbent (1973, p. 293).

[9] Quoted in Broadbent (1973, p. 257).

time when the idea that logic and imagination, as well as reason and feelings, represent worlds wide apart within the human mind, was commonly accepted.

By applying this distinction coherently, most rational design theories did not take into account the issue of *form* as priority. The layout contrived by the meta-design process could do as a sort of "generative cue" for the building plan. As for the building's morphology and appearance, the commonplace idea was that it should represent its purpose, complying the slogan "Form Follows Function".

2 Form and Function Between Reason and Nature

In a famous article written in 1896, "The Tall Office Building Artistically Considered", Louis Sullivan (1896) argued that contemporary American architects in their profession must face "something new under the sun" because a specific evolution and integration of social conditions resulted in the demand for a new typology of buildings: namely, tall office buildings were "a new grouping of social conditions [that] has found a habitation and a name".[10]

From this rational approach, Sullivan proceeded to explain the architectural nature of the problem: "How shall we impart to this sterile pile, this crude, harsh, brutal agglomeration, this stark, staring exclamation of eternal strife, the graciousness of those highest forms of sensibility and culture that rest on the lower and fierce passions? How shall we proclaim, from the dizzy height of this strange, weird, modern housetop, the peaceful evangel of sentiment, of beauty: the cult of a higher life?"[11]

Sullivan highlighted the architect's own task: that is, finding answers to questions that are both aesthetic and ethic. The problem of giving form to the habitation of this "new grouping of social conditions", that is, the modern office, is unprecedented. Therefore, in designing tall buildings, architects cannot resort to traditional rules, to the established "working tools" of the current profession. Instead, one should follow one's "natural instinct" and, after establishing the functional and technological structure of the tall building, one shall understand which parts of the building will need a special aesthetic connotation, within a harmonious overall composition, according to their own purpose and to their relationship with the city. Sullivan advocated "the erection of buildings finely shaped and charming in their sobriety", against any academic ornamentation, but his article has not the polemic tone and the dry wit of Adolf Loos's famous invective (Loos 1929)[12]. He rather includes decoration in the formal issue, which represents a higher order of enrichment, entailing a moral character and edifying aims. In fact, formal accomplishment allows the designer to advance the stage of the economic—functional programme, which left alone would produce "the sinister building of the speculator-engineer-builder combination". Once the material

[10] Sullivan (1896, p. 403).

[11] Ibid.

[12] A. Loos, *Ornament and Crime,* 1908.

aspects of the construction are resolved in the design draft, the architect must reason on the aspects concerning the spiritual nature, and therefore the feelings and emotions, that this kind of building should express and arouse. To accomplish this goal, architects should get rid of academic teaching. They should rather observe nature and consider the wonderful variety of natural forms: "All things in nature have a shape, that is to say, a form, an outward resemblance, that tells us what they are, that distinguishes them from ourselves and from each other. Unfailingly in nature these shapes express the inner life, the native quality, of the animal, tree, bird, fish, that they present to us; they are so characteristic, so recognizable, that we say, simply, it is natural. (…) Whether it be the sweeping eagle in his flight, or the open appleblossom, the toiling work-horse, the blithe swan, the branching oak, the winding stream at its base, the drifting clouds, over all the coursing sun, *form ever follows function, and this is the law*".

The three F's cliché "Form Follows Function" is an utter simplification which totally betrays Sullivan's ethical and poetic stance, but it became very popular in the mainstream culture of post-War architects and, apparently at least, in the practice of speculative building developments all over the world. Thus, it represented an easy target for the "anti-modern" reaction which burst out in the late 1960s. In his *Form Follows Fiasco,* author and architect Peter Blake (1977) proclaimed: "Most of the time the form is nothing but a probable hypothesis of the function. Most of the times in good (or more likely in bad) the form follows the current rates of the bank loan. Most of the times in modern architecture, the form is anti-functional. Most of the time these three assertions can be true".[13]

3 Good Fit, Permanence, Co-evolution: A Matter of Time

Rational design research was at its peak when Christopher Alexander published the work that gave him international fame (Alexander 1964). He was deeply involved in the search for a rational, formalized process in architectural design. In his research, he put *form* at the centre of the whole process: "The ultimate object of design is form". By this statement, he meant that any successful constructive process should result in a well-defined, well-shaped form, which necessarily responds to a number of environmental stresses, the way it happens in natural processes. So that there is no subjective judgement about what is good and bad because "good form" is the only possible one, the rational response to environmental forces.

In this, he referred to the studies of biologist and mathematician D'Arcy Wentworth Thompson, about how physical environmental forces shape the morphology of the living things in the course of their evolutionary growth (Thompson 1917). Alexander, a mathematician and an architect himself, underlined that D'Arcy W. Thompson even defined *form* as the *diagram of forces* for the irregularities that mark the relationship between living entities and their physical environment.

[13] Blake (1977, p. 40).

Following this line of thought, he explored the morphological development of human settlements according to the physical conditions that allow their birth and growth over time. He argued that a totally regular and homogeneous world would be completely amorphous, without any forces nor forms. Irregularities in the world's fabric are responsible for entities form, as *form* is the result of the world's entities efforts of *fitting* into an irregular environment. Accordingly, design cannot take into account a single object and its form alone, but "the ensemble comprising the form and its context". In fact, all design problems can be defined as "efforts to achieve *fitness* between two entities: the form in question and its context. The form is the solution to the problem; the context defines the problem". Alexander calls *good fit* the property of such an *ensemble*, which represents the design goal. In a famous example, he explains the problem of designing a traffic sign as the necessity of fitting the demands made on the sign by a driver's eye because "the ensemble is a truck driver plus a traffic sign".

Alexander's idea of function is far more complex than the representation given by the diagrams of system engineering, which shaped functional programmes in the rational design process and characterized the early phases of Systemics applied to architectural design. For him, *form* is the very focus of the problem.

"In a problem of design—Alexander argues—we want to satisfy the mutual demands which the two [elements of the ensemble] make on one another". How can we find the good fit for the ensemble of a human settlement plus its physical and social context? Our context is far too complex for a thorough operational description, "yet we certainly need a way of evaluating the fit of a form which does not rely on the experiment of actually trying the form out in the real world context. Trial-and-error design is an admirable method. But it is just real world trial and error which we are trying to replace by a symbolic method, *because real trial and error is too expensive and too slow*. (…)".[14]

Alexander looked for the formal rules of aggregation that could be abstracted by analysing "real life" human settlements and trying to translate their complex relationships into formal terms, using graphs and set theory, in order to discover their underlying order. His early efforts proved unsatisfactory and he quite early rejected some of this approach (Alexander 1965). Nevertheless, and in spite of this failure in defining a proper design method, his work brought into full light some very good points and questions: "Understanding the field of the context and inventing a form to fit are really two aspects of the same process. It is because the context is obscure that we cannot give a direct, fully coherent criterion for the fit we are trying to achieve (…) How is it, cognitively, that we experience the sensation of fit?". The consideration implies that we will never be able to make an exhaustive and finite list of positive requirements, which in real life represent a potentially infinite set. To approach the question, Alexander suggest a simple way of picking a finite set of requirements, by thinking of them in terms of *misfits*. He claims that it is easier to understand how and where a situation is not satisfactory: "This is because it is through misfit that the problem originally brings itself to our attention. We take just

[14] Alexander (1964, pp. 15–27).

those relations between form and context, which obtrude most strongly, which demand attention most clearly, which seem most likely to go wrong. *We cannot do better than this*"[15] This represents a sort of "fuzzy approach" towards the properties of good design: not a rigid list of requirement/performance prescriptions, which could never be very exhaustive, but rather a path of good advice against events "most likely to go wrong".

In "A City is not a Tree", Alexander recognizes that the vast majority of people, and a good number of architects as well, prefer old buildings and old cities to new ones. He calls new cities, deliberately planned and designed, *artificial cities*, while *natural cities* are "those cities which have arisen more or less spontaneously over many, many years".[16] He demonstrates that the formal organization of "natural cities" is a *semi-lattice*, the structure of living things where different activities can overlap and interact while belonging to different subsets, in opposition to the structure of "artificial cities", which can be represented like a *tree diagram*, where every subset separately stems like the branches of a tree. Alexander does not elaborate the issue of time; he just notes that also planned cities may become "natural" over time—like Liverpool and New York. In fact, a number of Roman towns had their origin as military camps, which is a typical tree organization, and nevertheless, "over many many years" they acquired the more subtle and more complex structure of a semi-lattice. Alexander does not openly indicate Time as one of the entities—or forces—that give the built environment its form. Nevertheless, when he writes that any living reality, any real system whose existence actually makes the city live, must be provided a physical receptacle, he implies that Time, flowing "over many many years" provides exactly the opportunity of physical receptacles for systems that had not been anticipated in the original plan.

Everything changes over time: Time, as a shaping force, destroys material things and overturns social structures—it breaks the boundaries that prevent overlapping. Alexander seems to admit that there is no possibility of planning a semi-lattice structure "because designers, limited as they must be by the capacity of the mind to form intuitively accessible structure, cannot achieve the complexity of a semi-lattice in a single mental act (…) for the human mind, the tree is the easiest vehicle for complex thoughts". Nevertheless, "the city is not, cannot and must not be a tree. A city is a receptacle for life (…) if we make cities which are trees, they will cut our life within to pieces".[17]

In the 1970s, important studies in the UK investigated the entities that give form to the built environment, focusing on flows of energy and matter which shape human settlements according to the local environmental characters (Martin and Steadman 1971; Martin and March 1972; Steadman 1975). These studies were intended to understand the urban morphogenesis in relation to the dynamics of the observed environmental variables and to develop operational models using topology applications, graph theory, functional interaction matrices, and other geometrical and mathematical techniques (Diappi 2004-2016; Broadbent 1973), thus providing "good design rules".

[15] Alexander (1965, pp. 58–62).

[16] Ibid.

[17] Ibid.

In Italy, architect and author Saverio Muratori (1959, 1963, 1967) investigated building typology and urban morphology according to their geographical, historical, and functional peculiarities. He researched the logic of morphogenesis in human settlements, with the main goal to provide a design tool for new developments in old cities. His *operante storia urbana* ("operational urban history") reconstructs the organic link between human groups and their human-made environment combining material history and geography theories along with intensive field work in Venice and in Rome. He studied typology and morphology in their material layering over time, in the process of constructing the built environment. Muratori recognized an "organic" relationship between human social groups and their settlements. Such relationship slowly produces physical and long-lasting transformations in the territory. In his view, the material construction of the built environment in subsequent, continuous layers within the mould of local geography, over time consolidates and selects morpho-typological characters, which condition later transformations and are in turn continuously transformed. Typology and morphology embody and express motifs which are formal, functional, cultural, and symbolic; their persistence is an indication of both adaptability and generative power. Muratori applied the results of his *operante storia urbana* in a project that won the competition for the CeP-Barene housing project in Venice Mestre (1960). Here, he designed the new development for public housing according to two alternative versions, using either the court or the linear type, both derived from the analysis of the historical Venetian built fabric.[18]

In this case, as in Alexander's, Time plays the role of the great Master Builder to which human settlements owe their most durable, *best fit* configuration. Thus, the durability of building forms appears to be an evolutionary quality, given the evolutionary nature of the urban phenomenon itself, whose dynamic morphology represents the material expression of flows of processes (Batty 2005; Marshall 2008; Batty and Marshall 2009).

4 Metaphors Aside: Feedback, Performance, Affordance

We can see that from early 20th Century to present days, many references to systemic thought can be found in the design field. Ludwig von Bertalanffy himself (von Bertalanffy 1968) wrote, as Sullivan and Alexander did, that every organic form is the expression of a flow of processes, persisting only in a continuous change of its

[18] Muratori's studies prompted typological studies by Giuseppe Caniggia (1981) and Pierluigi Cervellati (Cervellati and Scannavini 1973). Cervellati, an urban architect and town planning councillor for the municipality of Bologna from 1964 to 1980, was the promoter of the recovery of the historic centre of Bologna on the basis of the typological method, in an experience that had great international resonance. In the same years, building upon a completely different line of thought, Aldo Rossi (1966) investigated the "*logic of urban facts*" through the analysis of the cities' historical structure, in search of a *non-arbitrary* way for the construction of their future.

components, and that it can be considered as an expression of a pattern of processes of an orderly system of forces.

Currently, urban contexts are widely recognized as emerging and adaptive phenomena, inherently unstable because they exist in a continuous state of flux (Batty 2005).

A society's conceptual representation of its environment defines theories and tools to modify it. Thus, society promotes changes in the built environment according to the prevailing representation of environment within its "widespread culture". Italian geographer Eugenio Turri (1974) stressed the role of *systemic regulator* that culture can play: "As a biological ecosystem collapses when its use by the organism that inhabits it destroys its survival conditions, so its anthropological equivalent—the built environment—collapses when the balance between natural and human resources and the needs of its inhabitants are upset. (...) In this case, culture fails in its role of mediator between society and the environment, not being able to direct social behaviour and the actions of political and administrative institutions".

Today, unlike 40 years ago, it is perhaps easier for our society's widespread culture to accept the idea of the built environment as a specific and co-evolutionary eco-environment of the human species (Magnaghi 2000). Environmental failures and risks due to urban spread and overgrowth are all too obvious. On the other hand, the huge development of information technologies also help systemic references exit the metaphor, by developing methods and tools to analyse and model phenomena of increasing complexity. Architects should be aware that any design problems, as Alexander argued, should satisfy the mutual demands, which the two elements of the ensemble—design form and its environment—make on one another.

Actually, people and the built environment continuously exchange flows of matter, energy, and information. This should allow the idea of *feedback* properly entering the realm of architectural design.

Feedback regulates systems by integrating the information derived from the action-reaction circuits. In real-life built environment, this means studying its performance in relation to the needs, desires, and aspirations of the people who live and use it. For architects, this means learning from experience, recognizing and analysing mistakes and appreciating and disseminating success. Methods, techniques and tools have been developed over many years (Preiser et al. 1988) to analyse and evaluate the multi-faceted ways in which people react to buildings, spaces, and landscape. Performance-based evaluations developed increasingly their methods integrating ergonomics, proxemics, environmental psychology, anthropology, and sociology (Zeisel 2006; Preiser and Vischer 2005), which apply to different scales: buildings, urban neighbourhoods, landscapes (Mallory-Hill et al. 2012). Such sophisticated techniques make it possible to appreciate the relationship between people (individuals and social groups, their culture, expectations, social and economic conditions) and places (natural and built environment, resources, climate, use) in terms of physical and psychological perceptions, pleasure, satisfaction, preferences, and to express them in statistical form.

Performance-based approach to design may prove too "hard" and inadequate at wider space-and time-scales, when dealing with the ever-changing human built environment, which results from activities of generations over time. There is no fixed design system of controlling such flows, of eliminating uncertainty and flaws in their way.

Nevertheless, the relationship between the "users"—people, communities, human groups, and single beings—and the actions that shape our eco-techno evolutionary environment do require some kind of conceptualization to help appraisal for decision-making in the realm of common good.

A softer and promising approach to the ensemble "humans plus their physical and social context" seems to be the idea of *affordance,* coined by environmental psychologist J.J. Gibson (1979), who defined the *affordances* of the environment as the opportunities that it offers and provides to its inhabitants, according to their own specific characteristics. Affordance depends upon the physical properties of an environmental component, which are fit for the properties of an animal and allow the animal to use it in its ecological *niche.* Affordance characterizes the relationship between observer/user and its environment in terms of *opportunities* and involves cognitive, cultural, and social issues that are increasingly complex according to the species.[19]

The built environment in its development is subject to the shaping forces of human activities over time, with all the constraints and possibilities that Time and Nature put in its way. Over time, it becomes a goldmine of ever-changing *affordances.* The collective organizations of the human animal—communities—should be able to identify them to promote the species' survival.

References

Alexander, C. (1964). *Notes on the synthesis of form*. Cambridge: Harvard University Press.

Alexander, C. (1965). A city is not a tree. *Architectural Forum, 122*(1), 58–62.

Ashby, W. R. (1956). *An introduction to cybernetics*. London: Chapman & Hall.

Asimow, M. (1962). *Introduction to design*. Englewood Cliffs: Prentice-Hall.

Batty, M. (2005). *Cities and complexity*. Cambridge: MIT Press.

Batty, M., & Marshall, S. (2009). From Darwinism to planning-through Geddes and back. *Town and Country Planning, 78*, 462–464.

von Bertalanffy, L. (1968). *General system theory: Foundations, development, applications*. New York: Braziller.

Blake, P. (1977). *Form follows Fiasco: Why modern architecture hasn't worked.* New York: Little Brown & Company.

Boaga, G., & Giuffrè, R. (1975). *Metodo e progetto*. Roma: Officina.

Broadbent, G. (1973). *Design in architecture. Architecture and the human sciences*. London: Wiley.

Caniggia, G. (1981). *Strutture dello spazio antropico*. Firenze: Alinea.

Cervellati, P. L., & Scannavini, R. (1973). *Bologna: politica e metodologia del restauro nei centri storici*. Bologna: Il Mulino.

[19] In the design realm, the idea has been popularized by D. A. Norman, *The Psychology of Everyday Things*, New York, 1988, and simplified to define the degree of interaction between a designed object and its user.

Diappi, L. (Ed.). (2004-2016). *Evolving cities—Geocomputation in territorial planning*. New York: Routledge.
Gibson, J. J. (1979). *The ecological approach to visual perception*. Boston: Houghton Mifflin.
Giedion, S. (1941). *Space, time and architecture—The growth of a new tradition*. Cambridge: Harvard University Press.
Giedion, S. (1948). *Mechanization takes command: A contribution to anonymous history*. New York: Oxford University Press.
Habraken, N. J. (1998). *The structure of the ordinary—Form and control in the built environment*. Cambridge: MIT Press.
Hitchcock, H. R., & Johnson, P. (1932). *The international style*. New York: Norton & Company.
Le Corbusier. (1923). *Vers une Architecture*. Paris: G. Cres et Cie.
Loos, A. (1929). Ornament und Verbrechen, in Die Frankfurter Zeitung.
Magnaghi, A. (1973). *L'organizzazione del metaprogetto*. Milano: Franco Angeli.
Magnaghi, A. (2000). *Il progetto locale*. Torino: Bollati Boringhieri.
Mallory-Hill, S., Preiser, W. F. E., & Watson, C. (2012). *Enhancing building performance*. Oxford: Wiley-Blackwell.
Marshall, S. (2008). *Cities, design and evolution*. London: Routledge.
Martin, L., & March, L. (Eds.). (1972). *Urban space and structures*. Cambridge: Cambridge University Press.
Martin, L., & Steadman, P. (1971). *The geometry of environment: An introduction to spatial organization in design*. London: RIBA.
Muratori, S. (1959). *Studi per una operante storia urbana di Venezia*. Roma: Istituto Poligrafico dello Stato.
Muratori, S. (1963). *Studi per una operante storia urbana di Roma*. Roma: Consiglio Nazionale delle Ricerche.
Muratori, S. (1967). *Civiltà e territorio*. Roma: Centro studi di Storia e Urbanistica.
Norberg-Schulz, C. (1965). *Intentions in architecture*. Cambridge: MIT Press
Preiser, W. F. E., Rabinowitz, H. Z., & White, E. T. (1988). *Post occupancy evaluation*. New York: Van Nostrand Reinhold.
Preiser, W. F. E., & Vischer, J. C. (2005). *Assessing building performance*. Oxford: Elsevier.
Rossi, A. (1966). *L'architettura della città*. Padova: Marsilio.
Rudofsky, B. (1964). *Architecture without architects: A short introduction to non-pedigreed architecture*. New York: The Museum of Modern Art – Doubleday.
Rudofsky, B. (1977). *The prodigious builders—Notes towards a natural history of architecture with special regards to those species that are traditionally neglected of downright ignored*. New York: Harcourt Brace Jovanovich.
Steadman, P. (1975). *Energy, environment and building*. Cambridge: Cambridge University Press.
Sullivan, L. H. (1896, March). The Tall Office building artistically considered. *Lippincott's Magazine*, no. 57, pp. 403–409.
Thompson, D. W. (1917). *On growth and form*. Cambridge: Cambridge University Press.
Turri, E. (1974). *Antropologia del paesaggio*. Milano: Edizioni di Comunità.
Wiener, N. (1948). *Cybernetics: Control and communication in the animal and the machine*. Cambridge: MIT Press.
Zeisel, J. (2006). *Inquiry by design*. New York: Norton.

Climate Risks, Economics and Finance: Insights from Complex Systems

Francesco Lamperti, Irene Monasterolo, and Andrea Roventini

Abstract Climate change is posing daunting challenges to our societies. Such challenges are increasingly recognized by policymakers, practitioners and academics. Indeed, to limit the negative impact of human activities on the climate, 193 governments signed, in December 2015, the "Paris Agreement" aimed at stabilizing global temperature on 2 °C above pre-industrial levels. Meeting this goal requires massive private and public investments in low-carbon technological development, thus requiring a new role for the financial system and policymakers in the climate-finance nexus. Indeed, the financial system is expected to play a major role in shaping the speed, timing and pace of a sustainable transition by mobilizing capital. At the same time, policymakers ought to implement effective measures to foster new technologies and investments necessary to achieve a sustainable growth path. In addition, policymakers face the challenge to tackle the potential economic and financial risks associated to an uncoordinated low-carbon transition. In this chapter, we will employ a complexity perspective to study the risks, challenges and opportunities involved in the green transition, taking also into consideration the possible non-linearities, tipping points and path-dependency that characterize the co-evolution of climate, financial markets and economic dynamics.

F. Lamperti (✉) · A. Roventini
Institute of Economics, Scuola Superiore Sant'Anna, Pisa, Italy
e-mail: f.lamperti@santannapisa.it; francesco.lamperti@santannapisa.it; andrea.roventini@santannapisa.it

I. Monasterolo
University of Economics and Business (WU), Vienna, Austria
e-mail: irene.monasterolo@wu.ac.at

L. Urbani Ulivi (ed.), *The Systemic Turn in Human and Natural Sciences*, Contemporary Systems Thinking, https://doi.org/10.1007/978-3-030-00725-6_6

1 Introduction

In the last two decades, the understanding of the physical risks of climate change and their anthropogenic nature increased considerably (IPCC 2013; Hansen and Stone 2016). The fossil fuels based economic growth path, which took place since the industrial revolution, contributed to accumulate CO_2 emissions in the atmosphere at such a level that eventually triggered change in global temperatures and in the climate (Mann 2013). Climate change has a range of negative implications for ecosystems, economies and society that could irreversibly affect the quality of life of current and future generations (IPCC 2014; Stern 2006; Ackerman et al. 2014). Indeed, climate change is expected to increase the frequency of extreme weather events (e.g. floods, droughts) with heterogeneous impacts, yet leaving no-one excluded (Burke et al. 2015; Hallegatte and Rozenberg 2017) but hitting poor countries first (Schiermeier 2018). In particular, climate change is expected to affect poor households and vulnerable communities the most (Hallegatte et al. 2015), also in high-income countries (Hsiang et al. 2017), either because they are located in areas highly exposed (e.g. sea-level rise and coastal erosion in the Caribbean islands, Bueno et al. 2008) or because they are less able to cope with risk (Rhiney 2015). These impacts are associated to a degree of uncertainty that characterizes the precise localization and amount of climate damages at disaggregated scales, meaning that several estimates of losses could be even conservative.

These findings led scientists to call for immediate actions to decrease CO_2 emissions from human activities (a concept known as *mitigation*) and to invest in building resilience to climate change impacts at the global and local level (a concept known as *adaptation*, Solomon et al. 2007, 2014). At the UNFCCC COP21 conference that took place in Paris in 2015, most of UN countries signed the Paris Agreement aimed to limit global temperature increase below 2 °C on pre-industrial levels (UNFCCC 2016), while global temperature already increased by 1 °C on pre-industrial levels (Wuebbles et al. 2017). Limiting global temperature increase to 2 °C would require keeping Greenhouse Gases (GHG) emissions concentration in the atmosphere at a level close to 450 parts per million (ppm) by 2050 (Rogelj et al. 2015). This, in turn, implies an urgent transition to a sustainable, carbon-neutral production and consumption patterns, and the phasing-out carbon-intense investments. Article 2, para 1(c) of the Paris agreement explicitly defines the goal of "Making finance flows consistent with a pathway towards low greenhouse gas emissions and climate-resilient development".

It is now widely recognized that the transition to sustainability would come at a cost for investors both in the real economy (e.g. companies along the value-chain of value creation) and in finance. A fast and mobilization of capital in low-carbon sectors, in particular in the production and use of renewable energy, is fundamental to promote sustainable infrastructure development for the 9 billion (bn) people that would populate the planet by 2050 (NCE 2016). It has been estimated that investments required in the energy sector alone value an average $ 1 trillion (tn) circa per year by 2035 at the global level (IEA 2017), and Eur 180 bn per year in the European

Union (EU) alone (HLEG 2017). However, investments are not flowing at the pace and amount needed (CPI 2017), locking-in the economy and financial system into carbon-intense assets that are at risk of losing value due to climate change, the so-called carbon stranded assets (Leaton 2003). It has been estimated that to reach the 2 °C target, a large portion of reserves of oil, gas and coal should remain in the ground (McGlade and Ekins 2015) and will become *unburnable* (Leaton 2012; Pfeiffer et al. 2018). A recent study has evaluated in \$1300 bn the amount of stranded assets already present in the fossil fuel sector alone, and in \$25,000 bn the fossil fuels' built assets value that will be stranded by 2100 (Carbon Tracker 2018).

One of the barriers for shifting investments' allocation to green sectors and assets is the poor understanding of the relation between climate risks, the economy and finance. On the one hand, carbon-intensive investments contribute to increase GHG emissions concentration and thus climate change. On the other hand, they are at risk of loosing value via two channels: (1) climate physical risk (i.e. the destruction of immobilized capital as a consequence of extreme weather events and hazards) and (2) climate transition risks, linked to a delayed and uncoordinated transition to a low-carbon economy (Carney 2015). Due to the financialization of the economy that took place in the last three decades (Palley 2016), a loss in profits for companies exposed to stranded assets would cause a loss in the value of financial assets invested in those companies, namely equity, loans and bonds (Dietz et al. 2016; Battiston et al. 2017). In addition, the structure of tight interconnections of financial actors disclosed by after the break-out of the last financial crisis (Battiston et al. 2016b) means that climate risks could be amplified via reverberation among financial actors, giving rise to potentially systemic effects (ESRB 2016).

Thus, assessing the impact of climate change in its complexity and non-linearity, looking not only at the probability of future GDP losses but also at the spillover effects generated at the micro-level on the interconnected agents of the economy and finance, and their feedbacks on the stability of the overall system, is fundamental (see also Chichilnisky 2011). As recently recognized by Lord Stern, "many economic models add further gross underassessment of risk because the assumptions built into the economic modelling on growth, damages and risks, come close to assuming directly that the impacts and costs will be modest and close to excluding the possibility of catastrophic outcomes" (Stern 2013).

There is a growing literature that contributes to identify the research challenges and opportunities ahead for climate models (Burke et al. 2016; Diaz and Moore 2017). Nevertheless, the urgency to introduce effective policies and regulations to foster the transition of the economy to sustainability, and the need to align the financial system to this challenge (UNEP-FI 2018) calls for the development of alternative approaches for the analysis of the climate-economy-finance nexus.

At this regard, research recently discussed the advantages of approaches rooted on complexity science (Battiston et al. 2016a). In particular, Agent-Based Models (ABM), Stock-Flow Consistent models (SFC) and network models have flourished in the last decade also in response to the inability of traditional economic and financial pricing models to understand the reason of the last financial crisis, and in particular to assess the direct and indirect effects of endogenously generated shocks in contexts of market failures, imperfect information and the departure from full rationality and optimization. They have desirable features that allow them to consider the characteristics of climate impacts, i.e. non-linearity, time-delays and the presence of tipping points, in relation to the characteristics of the agents and sectors of real economy and finance. By relaxing strong assumptions on linearity and agents' rationality, such approaches allow to consider conditions of limited information and unbounded rationality (Farmer et al. 2015; Mercure et al. 2016; Balint et al. 2017), to overcome the limits of equilibrium assumptions by exploring the role of emerging properties, feedback loops and business cycles, as well as to consider the role of finance and financial interconnectedness in relation to the economy and climate. These features are important to provide a comprehensive and more realistic assessment of the relations along the climate-economy-finance nexus, and analyse key related policy issues such as the drivers of inequality and financial instability (Rezai and Stagl 2016). Therefore, they are best place to play a key role in the future of macroeconomics (Blanchard 2018; Stiglitz 2018) as well as to assess the economic and financial impacts of climate change.

Nevertheless, a reasoned overview on the recent methodological advances and contribution to the literature in this field is still missing. This chapter contributes to fill in this gap by presenting the rationale for the need of the alternative approaches to assess the economic and financial impacts of climate change. Then, it discusses to what extent evolutionary economics and complex system approaches could help disentangling the climate-economy and the climate-finance dimensions, by building on recent examples from the literature. The chapter is organized as follows. Section "Climate Change Economics: Why Current Models Are Wrong?" presents the characteristics of climate risk and uncertainty, discussing why traditional economic models and cost–benefit analyses are not a proper fit to assess climate impacts. Section "An Evolutionary Perspective on the Climate-Economy Dimension" reviews the contribution of evolutionary economics approaches to the analysis of the economic impacts of climate change, focusing on agent based models and SFC models. Section "A Financial Networks Perspective on Climate-Finance" discusses recent advances in the application of network models to the analysis of climate-related financial risks. Section "Conclusions" concludes providing recommendations for future research.

2 Climate Change Economics: Why Current Models Are Wrong?

One of the main barriers perceived by decision makers to delay action on climate change, either in terms of the introduction of climate policies or portfolios' divesting from carbon-intensive assets, is represented by the degree of uncertainty that characterizes the estimates of the socio-economic impacts of climate change (Burke et al. 2015; Revesz et al. 2014). Indeed, climate change cannot be considered as just another type of exogenous shock in relation to the economy and finance for several reasons:

- Climate change impacts (extreme weather events, e.g. floods or hurricanes) are characterized by fat tails and thus likely fit a power law rather than a normal distribution, and are expected to occur in the long term (i.e. after 2050) (Ackerman 2017).
- The presence of feedbacks that can either amplify (positive feedback) or diminish (negative feedback) the effects of climate forcing on initial warming. Reinforcing feedback mechanisms are particularly interesting in so far they often give rise to problems of path-dependency (Sterman 2000).
- The presence of tipping points, beyond which the characteristics of the system could dramatically change, contributes to increase the uncertainty of climate risk and impacts. For instance, global warming and permafrost thaws could lead the deposits of frozen methane and carbon dioxide lying beneath permafrost in Arctic regions, to be released into the atmosphere, with the risk of runaway warming (Vaks et al. 2013).

In addition, there has been growing discussion around the circularity between the evidence of climate impacts [e.g. extreme weather events, such as the Harvey flood (Mann et al. 2017[1])] and climate policies' introduction, as well as on the credibility of climate policies and investors' response in the economy and finance (Battiston et al. 2017). Finally, a main source of uncertainty, which is specific to the assessment of climate impacts on GDP, comes from the cost–benefit analyses traditionally used, and in particular in relation to the definition of the damage function and the social cost of carbon (Weitzman 2009, 2012; Nordhaus and Moffat 2017; Pindyck 2015; Ackerman and Stanton 2012).

Indeed, the choice of how to represent global warming-induced damages is the most speculative element of the analysis. Notwithstanding the burgeoning econometric literature on the assessment of the economic effects of climate change (see Carleton and Hsiang 2016 for a survey), selecting an appropriate functional form for the damages is still challenging. In many cases, models simply assess the impact of climate change on the economy via aggregate fractional GDP losses. The usual practice consists in specifying an ad-hoc functional form for the so-called damage function that expresses the percentage of output loss for any level of temperature anomaly. For example, Nordhaus (2008) uses an inverse quadratic loss function, Weitzman (2009) proposes a negative exponential functional specification that emphasizes the catastrophic role of large climate changes, while Tol (2009) uses sector- and area-specific loss functions.

The adoption of simple aggregate damage functions brings three issues. First, by considering only GDP losses, it is not possible to distinguish between different types of damage. Second, the adoption of continuous and "smooth" damage functions rule out the treatment of catastrophic, more or less rare, climate events and

[1] https://www.theguardian.com/commentisfree/2017/aug/28/climate-change-hurricane-harvey-more-deadly.

impose not to look at the fluctuations such events might create in the economic system. Finally, there is an absolute degree of certainty in the occurrence of the damage. This means that, whenever an increase in average surface temperature materializes, some output is deterministically destroyed. These issues result in a large inter-model variability in terms of expected damages from climate change that, in turns, translate into uncertainty with respect to the estimates of the social cost of carbon, i.e. the economic costs due to global warming under some emission mitigation policy (see Gillingham et al. 2015). Finally, we remark that additional variability is brought about by the choice of the utility function that Integrated Assessment Models (IAM) usually adopt[2]. As a consequence, the evaluation of climate impacts and the prevention strategy to be adopted is highly influenced by the choice of few parameters, whose value is extremely difficult to estimate. As suggested by a recent stream of studies (see section "An evolutionary perspective on the climate-economy dimension" for details), the contribution of IAM's evaluations of climate risks to be, in practice, close to useless or even dangerous both for policy design and impact assessment. A slightly more agnostic solution, where models are employed to robustly test alternative interventions in well-specified scenarios, and damages are qualitatively evaluated in terms of alternative long-run trajectories of the economy rather than in quantitatively poor GDP shares might be insightful and desirable.

Indeed, it emerges that in order to assess the impact of climate change on the economy and finance, it is fundamental to consider elements of complexity and non-linearity, as well as the cost of inaction, and to account for both the positive and negative externalities of climate mitigation and adaptation policies.

When it comes to socio-economic modelling, this implies at least two innovations on the state of the art. First, we need to look not only at the probability of future GDP losses but also at the spillover effects generated at the micro-level, on the interconnected network of agents of the economy and finance. Second, we need to understand to what extent could climate risks be amplified in a system of highly interconnected economic and financial agents, and the risk transmission channels on the stability of the overall system (Chichilnisky 2011; Sterman 2000; Meadows 2008). This compels us to move beyond the current sector-based approach to economic and financial risk analysis, and embrace a complexity and system-based approach.

Traditional economic models used to assess the impact of climate change and climate policies (i.e. the IAMs) have been widely criticized for not being able to capture and model these dimensions (Ackerman et al. 2009; Ackerman 2017; Stern 2016), thus providing unreliable results to policymakers (Stoerk et al. 2018).

The main limitations of traditional approaches pertain the use of strong assumptions on agents' representativeness, rationality, coordination and intertemporal optimization, as well as on the presence of a single equilibrium, reached via market-

[2]We underline that when referring to IAMs in this chapter we focus on Benefit-cost models, which is a subgroup of all integrated assessment models but the most represented in the economics literature.

clearing prices. Then, they tend to limit their analysis to the direct effect of a policy on the specific institutional sector targeted by the policy itself (e.g. banks, firms), thus neglecting feedback loops and underestimating its overall effect. Moreover, they consider the introduction of climate policies in terms of costs, neglecting the positive externalities and co-benefits of climate mitigation and adaptation. Further, they assume a quite infinite discount rate for effects occurring in the medium to long term. Finally, they don't include the financial sector characterized by interconnected agents, and its relation to the agents of the real economy. As a consequence, they misestimate the social cost of carbon as well as the potential systemic effects of climate shocks on the finance and from here to the real economy, and their distributive implications.

3 An Evolutionary Perspective on the Climate–Economy Dimension

On the climate–economics side, existing estimates of the potential impacts (in terms of both gains and losses) of increased climate change have been heavily criticized due to the characteristics (and limitations) of the models used, in particular as regard the assumptions on the discount rate, the chosen damage function and the computation of the social cost of carbon (see e.g. Ackerman et al. 2009; Ackerman and Stanton 2012; Pindyck 2013; Metcalf and Stock 2017).

Given the foregoing issues, one of the main advantages of complex system approaches (i.e. ABM, SFC and network models[3]) is to allow for a micro-level representation of the interactions between climate change and economic dynamics [as emphasized in particular by Moss (2000) and, more recently, Farmer et al. (2015) and Balint et al. (2017)] and the feedback loops occurring in the climate system, in the economy and—possibly—in their interactions.

Research on complex systems provides models and methods that are able to analyse the mechanisms of shocks' propagation and amplification that are key both to estimate catastrophic failures and to large shifts in socio-economic regimes. They also provide tools for evaluating the resilience, vulnerability or adaptability of a system and, hence, to assess the adequacy of climate policies with respect to the objectives of avoiding (or reducing the probability of) catastrophic climate impacts and of fostering the transition to a carbon-free economy. However, it should be accepted that, in the highly non-linear and uncertain settings, it is completely illusory to expect precise quantitative estimates of risk and losses, and to expect that socio-economic systems can be controlled. They can at best be partially understood and influenced.

[3] To the purposes of the present essay, we narrow down the analysis of complex system models to ABM, SFC and network models, despite being well aware of the broader modeling arena with respect to those we consider here.

In that, an evolutionary ABM and SFC can better account for the out-of-equilibrium dynamics shifting the economy from a business-as-usual to a green growth path, where a new stationary state could also not be reachable for years. Moreover, network models, SFC and ABM can provide a more accurate representation of climate-related damages considering distributional issues and the role of system connectivity. Relatedly, complexity-based models can be employed to study how climate change risks' impact on financial market dynamics. Finally, the fast pace at which ABM have blossoming in the last years has led to the development of a new generation of agent-based integrated assessment models (Lamperti et al. 2018c).

One of the first attempts to dynamically model a complex economy together with a climate module can be traced back to the LAGOM model family (Haas and Jaeger 2005). Heterogeneous households and producers face the risk of climate-related damages and are offered insurance contracts. An "expectation manager" helps insurers and households to update their expectations on the basis of new observations. LAGOM operates at multiple time scales: market interchanges occur much faster than climate change, and industrial production takes place at intermediate frequencies. The flexible accounting for different time scales is an advantage of ABMs vis-a-vis traditional IAMs, which usually consider yearly equilibrium adjustments both in the economic and climate system. Mandel et al. (2009) and, more recently, Wolf et al. (2013) have further extended the LAGOM model to simulate a growing economy with the possibility of specifying different interacting economic areas and to study the properties of economic growth as emerging from spatially explicit production networks. In each region, energy is produced within specific sectors with carbon emissions as a by-product. The model could then be used to test different mitigation policies.

Economic dynamics mainly affects climate change via the degree of environmental friendliness of production technologies, i.e. the amount of GHG emissions stemming from production. In general, production might involve goods, capital and energy. There are few sufficiently sophisticated agent-based models to deal with all these three aspects. For example, Beckenbach and Briegel (2010) limit themselves to the study of a generic production process, which is decomposed across different but not well-specified sectors. In a Schumpeterian setting, growth is triggered by firms' innovation and imitation strategies, and emission dynamics depends on two exogenous parameters governing the diffusion of low-carbon innovations and their quality.

Gerst et al. (2013) propose an agent-based model that endogenizes the process of technical change leading to the diffusion of less emission-intense machines. Drawing on the Keynes+Schumpeter model (Dosi et al. 2010), they study a complex economy composed of two vertically related industrial sectors and an energy production module, where competing technologies can be used to generate energy that is subsequently distributed through the system. The model is calibrated on US macroeconomic data and simulated until the end of the century to study different carbon tax recycling schemes. They find that only a policy focused on subsidies to carbon-free technology-oriented R&D allows a swift transition away from "dirty" energy technologies, and, in turns, to higher economic growth.

The major issues addressed in the contributions described so far is the identification of possible growth trajectories for both the economy and aggregate emissions, and in the adoption of fiscal policy (mainly carbon-taxes and subsidies) to direct the system towards some of these directions. The value added consists in the analysis of growth as a stable phenomenon emerging from an ecology of heterogeneous agents, whose different reactions to policies and uncertain environments can move the economy along trajectories that cannot be deduced otherwise. However, a key element is missing the picture. Indeed, the relationship between macroeconomic properties and the climate is explored in a single direction. The feedbacks that agents (firms, energy-production plants, households, etc.) receive from an increasing and possibly more volatile temperature have been generally ignored. Isley et al. (2013) construct a prototype for a hybrid agent-based integrated assessment model that could support the design of a government's regulatory climate policies. The authors underline the usefulness of the approach in analysing transformative solutions, that is, in examining how measures intended to reduce GHG emissions can trigger market-induced transformations, which, in turn, affect the government's ability to maintain its policy in an environment where agents affect the climate and receive back climate-related damages. However, in the latter framework, the climate system is left out of the picture and damages are linked to emissions, not to the average surface temperature. Moreover, environmental damages are modelled in a way that is akin to that embraced in standard IAMs (e.g. DICE, Nordhaus 1992) as aggregate cuts to potential GDP levels.

3.1 Evaluation of Climate Damages

In most integrated assessment models, climate damages are accounted for by an ad hoc damage function that impacts output (at the sectoral or the macro level) as a function of temperature increases brought about by GHG emissions. This approach ignores the propagation of shocks and the feedbacks that might relate damages to different sectors. Moreover, as most IAMs do not allow for agent heterogeneity, they entirely overlook distributional issues linked to climate damages.

Against this background, one of the characterizing features of complex systems lies in their representation of real phenomena as emerging from the interactions of heterogeneous agents. This approach allows to model the emergence of aggregate damages from micro shocks in production, procurement or finance percolating along network structures where households, firms, banks and the government interact. For instance, Hallegatte (2008) provides a model of shock propagation within Louisiana after the impact of hurricane Katrina. In the model, firms adapt their behaviour in an input-output network. Simulation results show that propagation mechanisms are essential for the assessment of the consequences of disasters, and that taking into account residual production capacities is necessary not to overestimate the positive economic effects of reconstruction. A straightforward consequence is the pivotal role played by the topology of the production network, which determines how firms are connected to each other and how (intermediate) goods flow though these links. Similarly, Henriet et al. (2012) disaggregate industry input-output tables to represent the production structure of regional economies at the firm level. They show that aggregate damages stemming from exogenous disasters are deeply affected by the network structure and the final outcomes depend especially on network concentration and clustering. In particular, concentration (degree of redundancy of suppliers and clients) acts as a risk sharing feature and clustering (degree of geographically dense interactions) allows small groups of interconnected firms to positively react to shocks happening outside the community they belong to.[4]

Systems' connectivity increases dramatically the complexity of studying the impact of climate events, and the impossibility to reduce the problem through simple aggregation or to impede failures at all scales calls for a re-design of how to model climate and weather damages (Helbing 2013).

Moving from a relatively restricted geographical focus to a global perspective, Bierkandt et al. (2014) introduce a model (labelled as acclimate) that is designed to evaluate the consequences of extreme climate events through the global supply chain. The model nests AB features (consumption and production sites are treated as agents) in an input-output network employed to track flows of goods in the system (taking also into account transportation). Acclimate is particularly well suited to study the propagation of shocks and it has been extended to better explore the differences between top-down cascades promoted by forwards linkages and demand-induced backward dynamics (Wenz et al. 2014). However, being the time scale of the model's simulations very short (from days to some weeks), price adjustment mechanisms are nearly absent, and technical change is overlooked. These shortcomings prevent from studying long-run macroeconomic dynamics. Nonetheless, this is a well-known problem in the literature: input-output models represent the short-term economic behaviour, in which production technologies are fixed and prices cannot adjust (Hallegatte 2014).

3.2 *Agent Based Integrated Assessment*

Despite the methodological advantages that agent-based models offer to the representation of production networks, the study of system's resilience and its reaction to different kind of shocks, there have been little efforts in employing these tools to investigate the effects of climate change on the aggregate economy. To the best of our knowledge, Lamperti et al. (2018a) introduce the first attempt to bridge a fully fledged agent-based integrated assessment model with a representation of climate-economic feedbacks, which take the form of stochastic shocks hitting agents with

---

[4] In particular, concentration (degree of redundancy of suppliers and clients) acts as a risk sharing feature and clustering (degree of geographically dense interactions) allows small groups of interconnected firms to positively react to shocks happening outside the community they belong to.

probability and size depending on the dynamics of the global mean surface temperature.

The model, called DSK, builds on Dosi et al. (2010, 2013) and is composed by two industries populated by heterogenous firms, a financial sector, an energy module, and a climate box grounded on Sterman et al. (2013). The model replicates a wide range of macro and micro stylized facts as well empirical regularities concerning climate change and economic dynamics (e.g. cointegration among energy consumption, GDP and GHG emissions). Given its satisfying explanatory power, the model can be employed as a laboratory to study the short- (transitions) and long-run (development trajectory) effects of a wide ensemble of climate, energy, innovation, fiscal and monetary policies. The model can also be extended to account for heterogeneous banks, financial markets and population growth. Lamperti et al. (2018a) show that micro-level climate impacts (whose average size is completely comparable with those in standard integrated assessment models) do not smooth out via the aggregation; rather, they result in economic dynamics which qualitatively differ from the business-as-usual and might also lead to stagnation, high volatility of economic fundamentals and surging unemployment. Exploiting the DSK model Lamperti et al. (2018b) have also shown that transitions to low carbon energy technologies are affected by the channels climate impacts hit the economic system and, further, that standard energy policy based on green incentives and carbon taxes might be largely ineffective in supporting a green transition under certain climate impact scenario, thereby pointing to the need of command and control (regulation based) policies.

3.3 *Macroeconomics, Climate and Finance*

The financing of the transition towards a low-carbon economy has still not been accurately explored in the economic literature. Indeed, as discussed above, the vast majority of modelling efforts focuses on government's fiscal policy. Recently, the role that financial and banking systems might play in inducing "green" investments and "green" entrepreneurship has received increasing attention (Campiglio 2016; Mazzucato 2015; Mazzucato and Penna 2016; Volz 2017). Different types of green fiscal (carbon tax, tax relief and breaks on investment in renewable energy) and targeted monetary policies (green bonds and quantitative easing) are simulated in the EIRIN model (Monasterolo and Raberto 2018a, b) which combines SFC and AB features. The model allows to represent heterogeneous financial and non-financial agents and sectors (including energy) as a network of interconnected balance sheet, where equilibrium conditions are replaced by balance sheet identities that hold true irrespective of any behavioural equations. The authors use the model to study the impact of green fiscal and green monetary policies, implemented via green sovereign bonds, and the phasing out of fossil fuels' subsidies, on green capital investment, credit market stability and income inequality. They find that the introduction of green policies, in particular those implemented via green bonds, has a green multiplier effect on the economy and doesn't imply instability in the credit and financial market. However, the distributive effects depend on the policies' design and implementation. Raberto et al.'s (2016) EURACE@UniGe model confirms such results. In such a context, the relation between fast de-carbonization policies and financial stability is emerging as a prominent concern on the climate policy agenda. On one side the financial system can foster the transition to a green development path. On the other side, it is increasingly exposed to climate risks.

Within this setting, the structure of the relationships among financial institutions might be crucial for the stability of the whole system. Focusing on this issue from a network perspective, Battiston et al. (2017) analyse the exposure of different classes of actors in the system using a well-known macro-network stress testing model (see also section "A Financial Networks Perspective on Climate-Finance"). They find that the direct exposure to fossil fuel and energy-intensive sectors, while limited overall, is relevant for investment funds, which in turns are highly connected with the banking system. Further, the housing sector can potentially trigger shocks which can be amplified by the financial system. Given the empirically well-documented degree of interdependences between actors in the financial, production and energy sides of the economy, the role of such relationships with respect to climate policy and their response to a changing climate is likely to be a challenge for future macro-oriented agent-based and network models.

4 A Financial Networks Perspective on Climate-Finance

The transition to a sustainable financial system is considered as fundamental to scale-up opportunities in the low-carbon transition and to avoid risks of carbon stranded assets (UNEP-FI 2018, HLEG 2017). However, designing a sustainable financial system that supports portfolios' divesting and shift to low-carbon investments requires a profound understanding of the relation between the drivers of climate-related risks, the channels of risk transmission and their impacts on the financial sector. At this regard, the development of metrics and methods to assess the exposure of individual financial portfolios (institutional actors) to climate physical and transition risks, as well as their contribution to climate action (i.e. mitigation and adaptation) has been identified as a main research topic by academics, practitioners and, recently, by financial regulators. On the one hand, financial network models have been developed and applied for stress testing portfolios of financial institutions against climate risks, thus introducing climate into financial risk assessment (e.g. the value-at-risk). On the other hand, financial macro-networks approaches have been developed to provide a comprehensive assessment of climate policies and risks, considering the feedback loops and the transmission channels within the financial and real economy sectors and agents.

4.1 Climate Risks and Financial Stability

There is growing awareness among central bankers and regulators of the potential negative implications of climate risks on prices and financial stability, and the need of tailored metrics and methods for portfolios' climate-related financial disclosure.

In 2015, the governor of the Bank of England, Mark Carney, addressed for the first time climate-related financial risks defining climate change as a "tragedy of the horizons" due to its intergenerational and distributive implications. In the speech he gave at Lloyds, he stated out that climate change effects will be mostly felt in the long term and would be suffered most by the current young and next generations who did not contribute to them. In 2016, the transmission channels from climate change to financial, i.e. climate physical, transition and liability risks, were identified in a report by the Bank of England (Batten et al. 2016). On the one hand, climate change could induce financial losses for the insurance and the banking sector as a result of climate-related events (such as droughts, hurricanes and floods). On the other hand, the transition to a low-carbon economy could lead to a re-pricing of carbon-intensive assets and thus to financial problems for companies whose revenues depend directly or indirectly on fossil fuels, with wider implications for financial stability. Several central banks followed, including Banque de France (Villeroy de Galhau 2015; Dombret 2018), the Bank of Italy (Signorini 2017), the Dutch Central Bank (Schotten et al. 2016) and most recently the director of the ECB Mario Draghi, who talked about the need to introduce climate and environmental factors in the assessment of price and financial stability (Draghi 2017). In addition, central banks started to consider the carbon risk assessment of their and financial actors' portfolios, and some already started to develop them (Regelink et al. 2017). The attention of central bankers increased after the release of the results of the first climate stress test of the financial system that showed a high exposure of financial actors' portfolios to carbon-intensive sectors (e.g. fossil fuels and fossil-based utility and transports), reaching even 45% for pension funds' equity holdings (Battiston et al. 2017). The research was based on a financial network model extended to account for climate policy shocks. More recently, a group of eight central banks and financial regulators have formed a "Network for Greening the Financial System", with the aim to develop metrics to supervise climate-related financial risks.

On the regulatory side, in 2016 the G20 Financial Stability Board (FSB) established a Task Force for Climate-related Financial Disclosures (TCFD) that provided in its final report sector-specific recommendations to foster companies' voluntary disclosure of climate-related financial risks. The FSB TCFD recommended the introduction of metrics and methods (e.g. climate stress test) to better inform their investors, lenders and insurance underwriters (see TCFD 2017). In 2017, the G20 launched a Task Force on "An International Financial Architecture for Stability and Development" that in 2018 recognized the importance of climate stress tests to scale up sustainable development finance.

In 2017, the European Commission launched the High-Level Experts Group on Sustainable Finance (HLEG) with the aim to provide recommendations to align the

European financial system to sustainability. In its final report (HLEG 2018), the HLEG focused on three main issues, i.e. the role of climate risk metrics for portfolios' disclosure, the introduction of a harmonized taxonomy for green bonds, and considerations on green macroprudential regulations in relation to banks' capital requirements.

4.2 Financial Networks for the Analysis of Climate-Related Financial Risks

Financial networks models and applications to stress testing and financial stability analysis based on the literature on complex networks and economic networks have been increasingly used to assess financial stability and provide insights on macroprudential regulation (Battiston and Martinez-Jaramillo 2018). Financial stress test based on network models has been increasingly applied by financial regulators in the aftermath of the 2008 financial crisis in their regular stress test activities to capture second-round effects in contagion exercises (BCBS 2015). The advantages of financial networks approach to stress test have been summed up by Battiston and Martinez-Jaramillo (2018) in the ability to:

- *Assess systemic risk* in case of relevant impact of indirect exposures on losses, diversification of risk across counterparties and external assets, and interconnectedness.
- *Understand the impact of externalities on financial contracts* and in the building up into systemic risk, considering the presence of imperfect information and incomplete risk markets.
- *Estimate the feedback effects* between the financial system and the real economy, where the network effects can better help to explain macro-economic aggregate phenomena.

More recently, the application of financial networks and stress test extended to consider the risk for financial stability induced by climate change. In particular, climate stress tests of investors' portfolios were recently indicated as a promising tool to enforce climate-related financial disclosure (Regelink et al. 2017; De Galhau 2018). Indeed, it has been highlighted that information gaps on portfolios' current exposure to carbon-intense assets and companies, and the lack of transparent metrics for climate-related financial disclosure represent major obstacles for portfolios' divesting in so far they prevent (1) investors to take climate into account in their portfolios' management strategies and credit risk assessment, (2) policymakers to introduce effective policies to smooth the low-carbon transition, and (3) central banks and regulators to assess the sources of risk for financial stability to inform their micro and macroprudential regulations. Recently, the contribution of network-based perspective to the analysis of carbon stranded assets (Campiglio et al. 2018) and of climate policies, and the implications for the development of robust carbon risk management policies for the global financial system and global supply chains (Mandel 2018) started to be analysed. Recent applications of network-based approaches to the analysis of climate-related financial risks led to the development of climate stress tests. These offer a framework to integrate climate consideration into financial and credit risk assessment, estimating individual portfolios' exposure to climate risks and impact of climate action (i.e. mitigation and adaptation). We can identify the following specific desirable features:

- The consideration of interconnectedness of financial portfolios that could amplify both positive and negative climate shocks and significantly decrease the accuracy of estimations of default probabilities, increasing the complexity of risk estimation.
- The extension of the time frame of the assessment of financial risk beyond the usual one of 1 year.
- The integration of climate considerations (e.g. physical and transition risk) into standard financial risk assessment and metrics, such as the value-at-risk.

The first network-based climate stress test of the financial system was developed by Battiston et al. (2017) and allows to assess the first and second round losses due to individual portfolios' exposure to climate-policy-relevant sectors, and to compute a climate Value-at-Risk (VaR) for an average brown or green bank's portfolio management strategy. The analysis also introduces a classification of economic activities at 4-digit NACE level accounting for their direct and indirect contribution to GHG emissions. The authors found that the value of the exposure of all the actors of the European financial system to carbon-intensive sectors that would be negatively affected by the introduction of climate policies (e.g. a carbon tax) is relevant, in particular for pension funds and investment funds, and reaches even 45–47% of their equity portfolios. Furthermore, the losses could be amplified by the mutual exposures of financial actors (e.g. pension funds and investment funds). These results show that climate-related financial disclosure is important for policymakers and regulators to identify who are the financial actors more exposed to climate risks. This, in turn, helps to introduce targeted policies aimed at fostering portfolios' shift to low-carbon investments and assets, and macroprudential regulations, thus decreasing the risk of systemic financial distress.

Climate stress test methodologies were also recently applied to the portfolios of development and national promotional banks. Monasterolo et al. (2018) developed the first carbon risk assessment for development banks applied to the overseas energy loans' portfolios of Chinese policy banks. The authors found that the Chinese policy banks' overseas energy portfolios are highly exposed to carbon-intense investments (in particular, fossil fuels-based projects) and to losses induced by climate transition risks. Potentially destabilizing effects would negatively impact both the borrower (i.e. the development bank) via the recovery rate in case of high leverage, and the beneficiary countries, via sovereign bonds' value. The analysis shows that negative shocks are mostly concentrated on coal and oil projects and vary across regions, model used to forecast the sectors' market-share shocks, and the climate policy scenario (i.e. milder or stricter). Shock ranges between 4.2% and 22% of total loans value, and the total value of losses of overseas energy loans of Chinese policy banks could reach ¼ of their total portfolio ($ 228 bn). Given the current leverage of Chinese policy banks, the authors conclude that these losses are not negligible in comparison to banks' capital.

The impacts of climate policy shocks have been also analysed in the context of financial macro-network models to estimate the propagation of distress in networks consisting of a set of financial or non-financial firms and set of financial contracts (equity holdings, loans, bonds) (Markose et al. 2017; Cimini et al. 2015; Battiston et al. 2016c). In a mark-to-market accounting environment, negative shocks induced by investors' unanticipated climate policies on equity values of firms could result in changes in the equities values of the other firms or financial actors holding their debt obligations (Battiston et al. 2016c; Bardoscia et al. 2017). Stolbova et al. (2018) developed a macro-financial network analysis for climate policy assessment, considering financial interconnectedness and closed chain of contracts, presence of bankruptcy costs, not fully collateralized debt contracts and recovery rate lower than one. Building on Castrén and Rancan (2014) and Battiston et al. (2016b), the authors build a methodological framework to assess the impacts of climate policies' shocks (including a carbon tax, a green quantitative easing, and green macroprudential regulation) to the balance-sheet of financial institutions through direct and indirect chains of financial exposures across multiple financial instruments. The financial macro-network approach is useful to assess the financial impact of climate risks for several reasons. First, when considering chains of closed contracts, we can identify the feedback loops that propagate climate shocks from a sector/agent to another one, and in particular the reinforcing feedback loops that could amplify the magnitude of the shock on the financial sector. Second, it allows us to analyse the implications of climate-related financial risks when we need to depart from the hypothesis of rational expectations. Indeed, it was shown that most market actors would not be able to anticipate the impact of technological shocks (e.g. the renewable energy revolution characterized by a fast decrease in cost and fast increase in productivity) and policy shocks (e.g. the USA withdrawing from the Paris Agreement) on assets' prices, thus leading to price volatility and systematic mispricing if systemic relevant asset classes and financial actors are involved (Monasterolo et al. 2017). In these conditions, the recovery rate on the obligations of all actors directly exposed to the asset that incurs losses can be significantly smaller than one. In a network of interconnected financial actors' balance sheets, where obligations of an actor are assets for another actor, there is risk of systematic mispricing of the value of the assets belonging to the second actor (Stolbova et al. 2018). The mispricing can then be propagated along the chain of financial contracts because in a mark-to-market accounting environment, market players make decisions based on the expected value of their counterparties' obligations (Battiston et al. 2016a; Bardoscia et al. 2017).

Therefore, the financial macro-network approach contributes to overcome the limits of traditional cost–benefit analyses of the economic impact of climate policies that overlook the financial sector and financial interconnectedness, and focus only on direct effects of policies on the specific institutional sector they target, thus neglecting possible feedbacks and underestimating the overall policy effect.

5 Conclusions

This chapter provides an overview of the approaches rooted on complexity science and evolutionary economics developed in the last decade to assess the economic and financial impact of climate change. This is a growing and promising field that could provide a breakthrough in the analysis of climate risks and opportunities for the economy and finance, to inform the timing and magnitude of climate policies. Indeed, these approaches have some desirable features that allow us to depart from the assumptions of market equilibrium, agents' intertemporal rationality and coordination of sector-based cost–benefit analyses and general equilibrium models, to account for the characteristics of complexity, non-linearity and path dependency of climate change and climate risks. Traditional approaches also don't consider monetary and financial variable, the relation between economic, financial and environmental dynamics, and neglect financial interconnectedness and complexity of risk, which is now recognized as a core element of financial markets. These are serious limitations that could lead to underestimate the overall impact of climate-related economic and financial risks, and in particular to underestimate both the costs of climate inaction (e.g. the negative externalities on human health and ecosystems) and the benefits of climate policies on building resilience to climate risks.

In particular, we focused on ABM and SFC models, network models and financial macro-network analyses. ABM and SFC models applied to the analysis of climate-related economic and financial impacts flourished in the last decade. The inability of traditional models to understand the last financial crisis, and to account for the characteristics of climate change and impacts, e.g. fat tails, non-linearity and tipping points, moved economists' attention to alternative approaches. ABM and SFC have desirable features when we want to model the relation between climate change, the economy and finance. Indeed, they allow modellers to depart from equilibrium conditions and from strong assumptions on agents' rationality, optimization and coordination, in order to model emerging behaviours. In addition, they embed money flows and a financial sector (despite with different level of details), sometimes connected to the energy sector and ecosystems.

Climate stress tests received growing attention by both academics and financial regulators for their contribution to understand the impact of both investors' portfolios direct and indirect exposures to carbon-intensive assets and sectors on financial stability, and the risk amplification due to financial interconnectedness. By introducing climate inside standard financial risk metrics, such as the value-at-risk, climate stress tests inform investors portfolios' management strategies and credit risk assessment.

Financial macro-network models that connect both financial and non-financial actors' balance sheets allow to estimate the channels of risk transmission from climate policies to finance and the real economy. In addition, they allow to identify the reinforcing feedback loops in the financial sector that could give rise to explosive effects, with negative implications on systemic risk spread and financial stability, and their cascade effects on the agents and sectors of the real economy.

Our review shows that complex systems science could offer flexible tools to analyse the relationship between the physical and the socio-economic system. By accounting for heterogeneous agents and their interactions, ABM, SFC and network models allow one to isolate mechanisms and effects that would otherwise be missing in the picture. This explains why these models have recently been put forward as prominent alternatives to standard models.

Nevertheless, further research advances are needed in order to improve the policy-relevance of these approaches. In particular, interdisciplinary research, as well as collaboration between academic research, central banks and regulators should be promoted. For instance, climate stress test could be a key tool for central banks to deliver on their mandate of price and financial stability, by considering the introduction of climate risks in their pricing and supervision tasks, as well as to assess whether green micro and macroprudential regulation is needed. Further, the financial macro-network approach informs policymakers on the potential unintended effects on financial risk and real economy contagion linked to the introduction of specific climate policies, that should be considered during the design and implementation of climate policies.

References

Ackerman, F. (2017). *Worst-case economics – extreme events in climate and finance*. London, New York: Anthem Press.

Ackerman, F., & Stanton, E. (2012). Climate risks and carbon prices: Revising the social cost of carbon. *The Open-Access, Open-Assessment E-Journal, 6*.

Ackerman, F., DeCanio, S. J., Howarth, R. B., & Sheeran, K. (2009). Limitations of integrated assessment models of climate change. *Climatic Change, 95*(3–4), 297–315. https://doi.org/10.1007/s10584-009-9570-x.

Ackerman, F., Stanton, E., Hope, C., & Alberth, S. (2014). Did the stern review underestimate US and global climate change? In F. Ackerman & E. Stanton (Eds.), *Climate change and global equity* (pp. 67–78). London: Anthem Press.

Balint, T., Lamperti, F., Mandel, A., Napoletano, M., Roventini, A., & Sapio, A. (2017). Complexity and the economics of climate change: A survey and a look forward. *Ecological Economics, 138*, 252–265.

Bardoscia, M., Battiston, S., Caccioli, F., & Caldarelli, G. (2017). Pathways towards instability in financial networks. *Nature Communications, 8*, 14416.

Batten, S., Sowerbutts, R., Tanaka, M. (2016). *Let's talk about the weather: The impact of climate change on central banks* (Staff Working Paper No. 603). Bank of England.

Battiston, S., & Martinez-Jaramillo, S. (2018). Financial networks and stress testing: Challenges and new research avenues for systemic risk analysis and financial stability implications. *Journal of Financial Stability, 35*, 6–16.

Battiston, S., Farmer, J. D., Flache, A., Garlaschelli, D., Haldane, A. G., Heesterbeek, H., Hommes, C., Jaeger, C., May, R., & Scheffer, M. (2016a). Complexity theory and financial regulation. *Science, 351*(6275), 818–819.

Battiston, S., D'Errico, M., & Gurciullo, S. (2016b). DebtRank and the network of leverage. *Journal of Alternative Investment, 18*(4), 68–81.

Battiston, S., Caldarelli, G., May, R. M., Roukny, T., & Stiglitz, J. E. (2016c). The price of complexity in financial networks. *Proceedings of the National Academy of Sciences, 113*, 10031–10036.

Battiston, S., Mandel, A., Monasterolo, I., Schütze, F., & Visentin, G. (2017). A climate stress-test of the financial system. *Nature Climate Change, 7*(4), 283–288. https://doi.org/10.1038/nclimate3255.

BCBS. (2015). *Making supervisory stress tests more macroprudential: Considering liquidity and solvency interactions and systemic risk*. BCBS Working Paper 29, Basel Committee for Banking Supervision.

Beckenbach, F., & Briegel, R. (2010). Multi-agent modeling of economic innovation dynamics and its implications for analyzing emission impacts. *International Economics and Economic Policy, 7*(2), 317–341.

Bierkandt, R., Wenz, L., Willner, S. N., & Levermann, A. (2014). Acclimate—A model for economic damage propagation. Part 1: Basic formulation of damage transfer within a global supply network and damage conserving dynamics. *Environment Systems and Decisions, 34*(4), 507–524.

Blanchard, O. (2018). On the future of macroeconomic models. *Oxford Review of Economic Policy, 34*(1–2), 43–54.

Bueno, R., Herzfeld, C., Stanton, E. A., & Ackerman, F. (2008). *The Caribbean and climate change*. Somerville, MA: Tufts University.

Burke, M., Dykema, J., Lobell, D. B., Miguel, E., & Satyanath, S. (2015). Incorporating climate uncertainty into estimates of climate change impacts. *The Review of Economics and Statistics, 97*(2), 461–471.

Burke, M., Craxton, M., Kolstad, C. D., Onda, C., Allcott, H., Baker, E., Barrage, L., Carson, R., Gillingham, K., Graff-Zivin, J., & Greenstone, M. (2016). Opportunities for advances in climate change economics. *Science, 352*(6283), 292–293.

Campiglio, E. (2016). Beyond carbon pricing: The role of banking and monetary policy in financing the transition to a low-carbon economy. *Ecological Economics, 121*, 220–230.

Campiglio, E., Godin, A., & Kemp-Benedict, E. (2018). Networks of stranded assets in a low-carbon transition. AFD Research Paper.

Carbon Tracker. (2018). *Mind the gap: The $1.6 trillion energy transition risk*. Report.

Carleton, T. A., & Hsiang, S. M. (2016). Social and economic impacts of climate. *Science, 353*(6304), p.aad9837.

Carney, M. (2015). *Breaking the tragedy of the horizon: Climate change and financial stability*. Bank of England.

Castrén, O., & Rancan, M. (2014). Macro-networks: An application to euro area financial accounts. *Journal of Banking & Finance, 46*, 43–58.

Chichilnisky, G. (2011). What is sustainability?. *International Journal of Sustainable Economy, 3*(2), 125–140.

Cimini, G., Squartini, T., Garlaschelli, D., Gabrielli, A., & Garlaschelli, D. (2015). Systemic risk analysis in reconstructed economic and financial networks. *Scientific Reports, 5*.

CPI, 2017. Global landscape of climate finance 2017. Climate Policy Initiative.

De Galhau, F. V. (2018). Green Finance – A New Frontier for the 21st Century. Speech.

Diaz, D., & Moore, F. (2017). Quantifying the economic risks of climate change. *Nature Climate Change, 7*(11), 774.

Dietz, S., Bowen, A., Dixon, C., & Gradwell, P. (2016). "Climate value at risk" of global financial assets. *Nature Climate Change, 6*(7), 676–679. https://doi.org/10.1038/nclimate2972.

Dombret, A. (2018). *Greener finance - better finance? How green should the financial world be? (speech delivered at the Bundesbank symposium "banking supervision in dialogue").* Frankfurt: Deutsche Bundesbank.

Dosi, G., Fagiolo, G., & Roventini, A. (2010). Schumpeter meeting Keynes: A policy-friendly model of endogenous growth and business cycles. *Journal of Economic Dynamics and Control, 34*(9), 1748–1767.

Dosi, G., Fagiolo, G., Napoletano, M., & Roventini, A. (2013). Income distribution, credit and fiscal policies in an agent-based Keynesian model. *Journal of Economic Dynamics and Control, 37*(8), 1598–1625.

Draghi, M. (2017). Response to a Letter of the Members of the European Parliament (L/MD/17/382).

ESRB. (2016). *Too late, too sudden: Transition to a low-carbon economy and systemic risk.* Advisory Scientific Committee Report No. 6. European Systemic Risk Board, Frankfurt.

Farmer, J. D., Hepburn, C., Mealy, P., & Teytelboym, A. (2015). A third wave in the economics of climate change. *Environmental and Resource Economics, 62*(2), 329–357.

Gerst, M. D., Wang, P., Roventini, A., Fagiolo, G., Dosi, G., Howarth, R. B., & Borsuk, M. E. (2013). Agent-based modeling of climate policy: An introduction to the ENGAGE multi-level model framework. *Environmental Modelling & Software, 44*, 62–75.

Gillingham, K., Nordhaus, W.D., Anthoff, D., Blanford, G., Bosetti, V., Christensen, P., McJeon, H., Reilly, J. and Sztorc, P. (2015). *Modeling uncertainty in climate change: A multi-model comparison* (No. w21637). National Bureau of Economic Research.

Haas, A., & Jaeger, C. (2005). Agents, bayes, and climatic risks-a modular modelling approach. *Advances in Geosciences, 4*, 3–7.

Hallegatte, S. (2008). An adaptive regional input-output model and its application to the assessment of the economic cost of katrina. *Risk Analysis, 28*(3), 779–799.

Hallegatte, S. (2014). Modeling the role of inventories and heterogeneity in the assessment of the economic costs of natural disasters. *Risk Analysis, 34*(1), 152–167.

Hallegatte, S., & Rozenberg, J. (2017). Climate change through a poverty lens. *Nature Climate Change, 7*(4), 250.

Hallegatte, S., et al. (2015). *Shock waves: Managing the impacts of climate change on poverty. Climate change and development.* Washington, DC: The World Bank. https://doi.org/10.1596/978-1-4648-0673-5.

Hansen, G., & Stone, D. (2016). Assessing the observed impact of anthropogenic climate change. *Nature Climate Change, 6*(5), 532.

Helbing, D. (2013). Globally networked risks and how to respond. *Nature, 497*(7447), 51–59.

Henriet, F., Hallegatte, S., & Tabourier, L. (2012). Firm-network characteristics and economic robustness to natural disasters. *Journal of Economic Dynamics and Control, 36*(1), 150–167.

High-Level Experts Group on Sustainable Finance (HLEG). (2017). Financing a sustainable European Economy. Interim Report.

High-Level Expert Group on Sustainable Finance (HLEG). (2018). Sustainable European Economy. Report to the European Commission, 1–100.

Hsiang, S., Kopp, R., Jina, A., Rising, J., Delgado, M., Mohan, S., Rasmussen, D. J., Muir-Wood, R., Wilson, P., Oppenheimer, M., & Larsen, K. (2017). Estimating economic damage from climate change in the United States. *Science, 356*(6345), 1362–1369.

IEA. (2017). Market report series: Renewables 2017. Analysis and Forecasts to 2022. Paris, 2017.

IPCC. (2013). Summary for policymakers. In T. F. Stocker, D. Qin, G.-K. Plattner, M. M. B. Tignor, S. K. Allen, et al. (Eds.), *Climate change 2013: The physical science basis. Contribution of working group I to the fifth assessment report of the intergovernmental panel on climate change*. Cambridge, UK, and New York: Cambridge University Press. http://www.climate2013.org/spm.

IPCC. (2014). Climate change working group III contribution to the IPCC fifth assessment report. In O. Edenhofer, R. Pichs-Madruga, Y. Sokona, E. Farahani, S. Kadner, K. Seyboth, A. Adler,

I. Baum, S. Brunner, P. Eickemeier, et al. (Eds.), *Climate change 2014: Mitigation of climate change*. Cambridge (UK) and New York (US): Cambridge University Press.
Isley, S., Lempert, R., Popper, S., & Vardavas, R. (2013). *An evolutionary model of industry transformation and the political sustainability of emission control policies*. Technical report, RAND Corporation - TR-1308-NSF.
Lamperti, F., Dosi, G., Napoletano, M., Roventini, A., & Sapio, A. (2018a). Faraway, so close: Coupled climate and economic dynamics in an agent-based integrated assessment model. *Ecological Economics, 150*, 315–339.
Lamperti, F., Dosi, G., Napoletano, M., Roventini, A., & Sapio, A. (2018b). *And then he wasn't a she: Climate change and green transitions in an agent-based integrated assessment model*. LEM Working Papers.
Lamperti, F., Mandel, A., Napoletano, M., Sapio, A., Roventini, A., Balint, T., & Khorenzhenko, I. (2018c). Towards agent-based integrated assessment models: Examples, challenges, and future developments. *Regional Environmental Change*, 1–16.
Leaton, J. (2012). *Unburnable carbon—Are the world's financial markets carrying a carbon bubble*. London: Carbon Tracker Initiative.
Leaton J. (2013). Unburnable carbon: Wasted capital and stranded assets. Carbon Tracker Initiative & Grantham Research Institute on Climate Change and the Environment.
Mandel, A. (2018). *Complex networks and climate policy*. Working paper.
Mandel, A., Fürst, S., Lass, W., Meissner, F., & Jaeger, C. (2009). Lagom generiC: An agent-based model of growing economies. In *European Climate Forum, Working Paper* (Vol. 1, p. 2009).
Mann, M. E. (2013). *The hockey stick and the climate wars: Dispatches from the front lines*. New York, NY: Columbia University Press.
Mann, M. E., Rahmstorf, S., Kornhuber, K., Steinman, B. A., Miller, S. K., & Coumou, D. (2017). Influence of anthropogenic climate change on planetary wave resonance and extreme weather events. *Scientific Reports, 7*, 45242.
Markose, S. M., Giansante, S., & Rais Shaghaghi, A. (2017). A systemic risk assessment of OTC derivatives reforms and skin-in-the-game for CCPs. *Financial Stability Review, 21*, 111–126.
Mazzucato, M. (2015). The (green?) entrepreneurial state. In I. Scoones, M. Leach, & P. Newell (Eds.), *The politics of green transformations*. New York: Routledge.
Mazzucato, M., & Penna, C. C. (2016). Beyond market failures: The market creating and shaping roles of state investment banks. *Journal of Economic Policy Reform, 19*(4), 305–326.
McGlade, C., & Ekins, P. (2015). The geographical distribution of fossil fuels unused when limiting global warming to 2C. *Nature, 517*, 187–190.
Meadows, D. H. (2008). *Thinking in systems: A primer*. White River Junction, VT: Chelsea Green Publishing.
Mercure, J. F., Pollitt, H., Bassi, A. M., Viñuales, J. E., & Edwards, N. R. (2016). Modelling complex systems of heterogeneous agents to better design sustainability transitions policy. *Global Environmental Change, 37*, 102–115.
Metcalf, G. E., & Stock, J. H. (2017). Integrated assessment models and the social cost of carbon: A review and assessment of US experience. *Review of Environmental Economics and Policy, 11*(1), 80–99.
Monasterolo, I., & Raberto, M. (2018a). *Is there a role for central banks in the low-carbon transition? A stock-flow consistent modelling approach*. Working paper presented at the CEP-DNB Workshop on "Central Banking and Green Finance", November 28–29, Amsterdam.
Monasterolo, I., & Raberto, M. (2018b). The EIRIN flow-of-funds behavioural model of green fiscal policies and green sovereign bonds. *Ecological Economics, 144*, 228–243.
Monasterolo, I., Battiston, S., Janetos, A. C., & Zheng, Z. (2017). Vulnerable yet relevant: The two dimensions of climate-related financial disclosure. *Climatic Change, 145*(3–4), 495–507.
Monasterolo, I., Zheng, J. I., & Battiston, S. (2018). Climate transition risk and development finance: A climate stress-test of China's overseas energy portfolios. Input to the G20 Task Force "An International Financial Architecture for Stability and Development", Forthcoming on China and the World Economy Journal.

Moss, S. (2000). Messy systems—The target for multi agent based simulation. In S. Moss & P. Davidsson (Eds.), *Multi agent based simulation, LNCS series* (Vol. 1979). New York: Springer.

New Climate Economy. (2016). *The sustainable infrastructure imperative: financing for better growth and development*. London, UK.

Nordhaus, W. D. (1992). An optimal transition path for controlling greenhouse gases. *Science, 258*(5086), 1315–1319.

Nordhaus, W. D. (2008). *A question of balance: Economic modeling of global warming*. New Haven: Yale University Press.

Nordhaus, W. D., & Moffat, A. (2017). *A survey of global impacts of climate change: Replication, survey methods, and a statistical analysis* (No. w23646). National Bureau of Economic Research.

Palley, T. (2016). *Financialization: The economics of finance capital domination*. Berlin, Heidelberg: Springer.

Pfeiffer, A., Hepburn, C., Vogt-Schilb, A., & Caldecott, B. (2018). Committed emissions from existing and planned power plants and asset stranding required to meet the Paris agreement. *Environmental Research Letters, 13*(5), 054019.

Pindyck, R. (2013). Climate change policy: What do the models tell us? *Journal of Economic Literature* Retrieved from http://www.ingentaconnect.com/content/aea/jel/2013/00000051/00000003/art00007, *51*, 860.

Pindyck, R. S. (2015). *The use and misuse of models for climate policy*. Working Paper 21097, National Bureau of Economic Research.

Raberto, M., Ozel, B., Ponta, L., Teglio, A., & Cincotti, S. (2016). *From financial instability to green finance: the role of banking and monetary policies in the Eurace model* (No. 2016/07).

Regelink, M., Reinders, H. J., Vleeschhouwer, M., & van de Wiel, I. (2017). *Waterproof? An exploration of climaterelated risks for the Dutch financial sector*. De Nederlandsche Bank.

Revesz, R. L., Howard, P. H., Arrow, K., Goulder, L. H., Kopp, R. E., Livermore, M. A., Oppenheimer, M., & Sterner, T. (2014). Global warming: Improve economic models of climate change. *Nature, 508*(7495), 173–175.

Rezai, A., & Stagl, S. (2016). Ecological macroeconomics: Introduction and review. *Ecological Economics, 121*, 181–185.

Rhiney, K. (2015). Geographies of Caribbean vulnerability in a changing climate: Issues and trends. *Geography Compass, 9*(3), 97–114. https://doi.org/10.1111/gec3.12199.

Rogelj, J., Luderer, G., Pietzcker, R. C., Kriegler, E., Schaeffer, M., Krey, V., & Riahi, K. (2015). Energy system transformations for limiting end-of-century warming to below 1.5 C. *Nature Climate Change, 5*(6), 519.

Schiermeier, Q. (2018). Clear signs of global warming will hit poorer countries first. *Nature, 556*(7702), 415.

Schotten, G., van Ewijk, S., Regelink, M., Dicou, D., & Kakes, J., 2016. *Time for Transition-an exploratory study of the transition to a carbon-neutral economy* (Occasional Studies No. 14–2). Netherlands Central Bank.

Signorini, L. F. (2017). *The financial system, environment and climate: A regulator's perspective (welcome address by the deputy governor of the Bank of Italy)*. Rome: Bank of Italy.

Solomon, S., Qin, D., Manning, M., Chen, Z., Marquis, M., Averyt, K.B., Tignor, M. & Miller, H. L. (2007). IPCC, 2007: Climate change 2007: The physical science basis. Contribution of Working Group I to the fourth assessment report of the Intergovernmental Panel on Climate Change. S. D. Solomon (Ed.).

Sterman, J. D. (2000). Business dynamics: Systems thinking and modeling for a complex world (No. HD30. 2 S7835 2000).

Sterman, J. D., Fiddaman, T., Franck, T., Jones, A., McCauley, S., Rice, P., Sawin, E., & Siegel, L. (2013). Management flight simulators to support climate negotiations. *Environmental Modelling & Software, 44*, 122–135 Thematic Issue on Innovative Approaches to Global Change Modelling.

Stern, N. (2006). The stern review on the economic effects of climate change (Report to the British Government). HM Treasury–Retrieved from www. sternreview. org. uk.

Stern, N. (2013). The structure of economic modeling of the potential impacts of climate change: Grafting gross underestimation of risk onto already narrow science models. *Journal of Economic Literature, 51*(3), 838–859.

Stern, N. (2016). Economics: Current climate models are grossly misleading. *Nature News, 530*(7591), 407.

Stiglitz, J. E. (2018). Where modern macroeconomics went wrong. *Oxford Review of Economic Policy, 34*(1–2), 70–106.

Stoerk, T., Wagner, G., & Ward, R. E. (2018). Recommendations for improving the treatment of risk and uncertainty in economic estimates of climate impacts in the sixth intergovernmental panel on climate change assessment report. *Review of Environmental Economics and Policy, 12*(2), 371–376.

Stolbova, V., Monasterolo, I., & Battiston, S. (2018). A financial macro-network approach to climate policy evaluation. *Ecological Economics, 149*, 239–253.

TCFD. (2017). Recommendations of the Task Force on Climate-related Financial Disclosures. Task Force on Climate-related Financial Disclosures.

Tol, R. S. (2009). The economic effects of climate change. *Journal of Economic Perspectives, 23*(2), 29–51.

UNEP-FI. (2018). *Making waves: Aligning the financial system with sustainable development.* Report.

UNFCCC. (2016). *Report of the Conference of the Parties on its twenty-first session, held in Paris from 30 November to 13 December 2015. Addendum. Part two: Action taken by the Conference of the Parties at its twenty-first session* (No. FCCC/CP/2015/10/Add 1). United Nations Framework Convention on Climate Change, Paris.

Vaks, A., Gutareva, O. S., Breitenbach, S. F., Avirmed, E., Mason, A. J., Thomas, A. L., et al. (2013). Speleothems reveal 500,000-year history of Siberian permafrost. *Science, 340*, 183–186. https://doi.org/10.1126/science.1228729.

Villeroy de Galhau, F. (2015). *Climate change: The financial sector and pathways to 2°C (speech by the governor of the Banque de France).* Paris: Banque de France.

Volz, U. (2017). On the role of central banks in enhancing green finance. UN Environment Inquiry/ CIGI Research Convening, United Nations Environment Programme, 2017.

Weitzman, M. L. (2009). On modeling and interpreting the economics of catastrophic climate change. *The Review of Economics and Statistics, 91*(1), 1–19.

Weitzman, M. L. (2012). GHG targets as insurance against catastrophic climate damages. *Journal of Public Economic Theory, 14*(2), 221–244.

Wenz, L., Willner, S. N., Bierkandt, R., & Levermann, A. (2014). Acclimate—A model for economic damage propagation. Part II: A dynamic formulation of the backward effects of disaster-induced production failures in the global supply network. *Environment Systems and Decisions, 34*(4), 525–539.

Wolf, S., Fürst, S., Mandel, A., Lass, W., Lincke, D., Pablo-Martí, F., & Jaeger, C. (2013). A multi-agent model of several economic regions. *Environmental Modelling & Software, 44*, 25–43.

Wuebbles, D. J., Fahey, D. W., Hibbard, K. A., DeAngelo, B., Doherty, S., Hayhoe, K., Horton, R., Kossin, J. P., Taylor, P. C., Waple, A. M., & Weaver, C. P. (2017). Executive summary. In D. J. Wuebbles, D. W. Fahey, K. A. Hibbard, D. J. Dokken, B. C. Stewart, & T. K. Maycock (Eds.), *Climate science special report: Fourth National Climate Assessment, volume I* (pp. 12–34). Washington, DC, USA: U.S. Global Change Research Program. https://doi.org/10.7930/J0DJ5CTG.

The Legal Concept of the Environment and Systemic Vision

Maurizio Cafagno, Domenico D'Orsogna, and Fabrizio Fracchia

Abstract Environment has not traditionally been considered by law in a systemic perspective. The work, after recalling the dynamics of legal evolution, tries to fill this gap. In particular, considering the characteristics of environmental resources and ecosystem services, it comes to propose a summary formula, according to which environmental objects point to law as commons with legacy value. The final part of the work uses climate change as test bench, concerning a typical example of global and complex problem.

The chapter is a reworking of a contribution presented jointly by the authors on 23 May 2014 at the Università Cattolica del Sacro Cuore of Milan, as part of the cycle of interdisciplinary seminars entitled "Sistemiche in filosofia, nelle scienze e nelle arti", now collected in L. URBANI ULIVI (editor), Strutture di mondo. Il pensiero sistemico come specchio di una realtà complessa, Il Mulino, 2010 (I), 2013 (II), 2015 (III). In the economy of shared effort and reflection, sections 1 and 2 of the contribution were written by Domenico D'Orsogna, sections 3 to 6 by Maurizio Cafagno, and sections 7 to 11 by Fabrizio Fracchia.

M. Cafagno (✉)
Dipartimento di diritto, economia e culture, Università degli studi dell'Insubria, Como e Varese, Italy
e-mail: maurizio.cafagno@uninsubria.it; maurizio@cafagno.com

D. D'Orsogna
Università di Sassari, Sassari, Italy
e-mail: dorsogna@uniss.it; ddo@studiolegaledorsogna.it

F. Fracchia
Università Bocconi, Milano, Italy
e-mail: fabrizio.fracchia@unibocconi.it

L. Urbani Ulivi (ed.), *The Systemic Turn in Human and Natural Sciences*, Contemporary Systems Thinking, https://doi.org/10.1007/978-3-030-00725-6_7

1 The Law and Its Environment

To position the issue of environmental protection from a legal point of view, it is worth examining, on one side, the way in which the environment, as a reality, is considered by the law; and, on the other, the responses that the law offers, in terms of regulation, to the need to protect interests related to the environment.

As regards the first aspect, it is not unreasonable to state that the legal notion of the environment has long been understood as being the exact opposite of a "systemic" notion,[1] in sharp contrast to the indications coming from the "non-legal" sciences (first and foremost, ecological science), which, instead, have long viewed the environment as a complex, dynamic, continuous, adaptive, resilient, productive "system" of (ecosystemic) services.[2]

Only recently, as will be explained later, this basic notion has entered the legal sphere as well (Cafagno 2007); but, to date, it cannot be said that this represents a notion which is shared to the extent that it might constitute the paradigm of reference for the dominant legal approach (theoretical and of positive law) to the issue of the environment.[3]

---

[1]About the notion of "system" in the legal sphere, cf. above all Falzea (1996); as well as the extensive investigation by Mario G. Losano, in three volumes (2002), which also critically examines the profound differences that can be seen between the "systematic" approach (in its many historical or theoretical variations), traditionally practised in the legal field, and the "systemic" view of law, as elaborated by Luhmann and Teubner. According to Losano, "the legal system proposed by Luhmann is to be found in a perspective which is not only completely different from that of jurists, but also to a large extent foreign to them [...]: the legal system is seen as a system of social reality, as a subsystem of society [...] and legal science passes from a system of concepts to a system of actions [...]; it completely changes the perspective from which one looks at the law: one looks at it from the perspective of the sociologist, and not from that of the law. Therefore, the relationship between the notion of a system (however it is defined) and that of law (however it is defined) also changes completely. Until Luhmann the system in law had been studied; with Luhmann one studies law in the system [ibid, 337–338, 347–348]. The best known "systematic" conception of twentieth-century law is, however, as noted, that of Hans Kelsen, the influence of whose theses is responsible for the shift of interest recorded in doctrine, starting from the 1960s, from the norm to the legal system, from fragmentation to the "systematic nature" of the law. See also Benvenuti (1982), Gomarasca (2010), Romano (1946), Troper (2001).

[2]Cf. Cafagno (2007) to whom for the moment reference is also made for complete bibliographical indications on the topics that will be dealt with in the first two sections.

[3]Environmental law is the subject of a vast doctrinal output, which cannot be fully described here. We therefore limit ourselves to mentioning, without any pretence of completeness, some general works: Grassi et al. (1999), De Carolis et al. (2006), Ferrara (2000, 2006), Rapisarda Sassoon (2002, 34 ff.), di Plinio and Fimjani (2002), Cugurra et al. (2006), Bucello and Cafagno (2005), Capaccioli and Dal Piaz (1980, I, 257 ff.), Caravita (2005), Cecchetti (2000), Cioffi (2009, 970), Civitarese Matteucci (1992, 662; 2003, 253 ff.), Cordini (2012), Crosetti et al. (2002), D'Amelio (1988), Dell'Anno (2003), Fonderico (2003, 2084 ff.; 2008), Fracchia (2002, 215 ff.; 2013), Giannini (1973, 15 ff.), Grassi (2003, 979 ff.), Maddalena (1990, 469 ff.), Mezzetti (2002), Morbidelli (1996, 1121 ff.), Pericu (1987, 189 ff.), Porena (2009), Postiglione (1985, 32 ff.), Pototschnig (1970, 459 ff.), Predieri (1981, 503 ff.), Renna (2012, 62 ff.), Scoca (1993, 399 ff.), Tallacchini (1996).

The problem of environmental protection in the legal field undoubtedly recalls a notion of the environment derived from non-legal sciences, which the lawyer uses as their "working materials".[4]

It is useful to state, furthermore, that the law does not protect the environment as such (in and of itself, as an object to be safeguarded), but it takes it into account to the degree in which its characteristics assume a value functional to the life of humanity and its needs.

Environmental protection, understood in this way, is homogeneous with respect to the purposes of the legal system.

In reality, only those facts that have an impact on interests are of concern to the law; the others are extraneous to the interest of humanity and, consequently, to the law.

The legal phenomenon, considered in its complexity and dynamism, does indeed tend to provide order to the social reality, external to it, regulating what is between (*inter esse*) humanity and assets: interests.

The regulation of interests (generated by facts, understood in the general sense of natural events and human behaviour) is the specific objective of the law; the law finds its complete expression in this and, with this, its limits.

The law is a social science that deals with the problems of humanity as a *civis* and, therefore, it is possible to deal with the environment with constant reference to humanity.

In the ethical or philosophical field, there may well be theories that consider the relationship between humanity and the environment in different ways, in which humanity is emphasised in a more or less accentuated manner or seen as a mere element, like the others, of the environment and the ecosystem.[5]

But the theses which, in the legal sphere, propose to move (from an anthropocentric vision of the law) in a biocentric or ecocentric direction seem to be, on closer inspection, "out of tune" with respect to the logical horizon of the law (Scoca 1993, 399 ff.; Habermas, 1998).

The evolution of environmental law signals, if anything, an extension of the field (but also an axiological reorientation) of the anthropocentric view of law, extended to the consideration of the interests of humanity as a species and of intergenerational equity.[6]

---

[4] Cf. Sorace (1999, 125): "Jurists must not delude themselves that they can do their job by knowing only the law. They must first know the phenomena that they will have to deal with employing the tools of their profession. Naturally, the in-depth knowledge that only other specialists may have is not required but they have to have some basic knowledge of the phenomena of the metalegal reality on which they have to intervene and, above all, be able to dialogue with other specialists, thus grasping all the aspects of the phenomena that are legally relevant. It is therefore evident that dealing with the environment is a particularly demanding challenge for jurists"; in the same sense, De Leonardis (2005, 202, spec. note 68). The need to be open to an interdisciplinary approach is indicated by, among others, Caravita (2005, 22 ff.), Cordini (1994).

[5] The literature on this subject is vast. Cf., *ex multis*, Habermas (1996), Iovino (2004), Bocchi and Ceruti (2004, 169 ff.), Tallacchini (1996, 16 ff.), Zito (2006, 3 ff.).

[6] It is wise to underline that "the attribute of the sustainability imposed on development prescribes

The environment, therefore, assumes importance for the law to the extent, which is gradually widening, in which it encounters the need to protect human interests.[7]

As regards the second profile of analysis set out at the start (the responses that the law offers to the protection of environmental interests in terms of regulation), it is useful to underline how the environment poses a series of peculiar problems for legal regulation: the need to manage complexity and a lack of information, technological uncertainty, the need for flexibility, adaptation, and reversibility of choices, downscaling, coordination, participation, and information of the (public and private) players involved, for proper communication between legal instruments and economic market incentives (Fracchia 2010), of the traverse, cross-border and inter-sectoral nature of the legal disciplines.

Some of these problems refer to a question of general theory of law: the problem posed by the material efficacy of (dynamic) facts which continually affect the material that is the subject of legal regulation, demanding a constant adjustment of the regulations to the ongoing dynamism and mutability of the fact.[8]

In other words, there are facts (endowed with a material dynamism) which pose peculiar problems for the law. These are facts that act continually as factors that generate (and modify) interests, and which therefore require an ongoing adaptation of the legal discipline that relates to them (*rectius*: that relates to the interests originating from them).

The environment (however, it is considered or described by the norms) comes into this category. With regard to its protection, it is useful to recall the so-called principle of "convenience" (or, if we prefer, congruence) of the legal effect on the fact, on the basis of which the legal framework established by the law has to offer a coherent response to the problem (of regulation of interests) posed by the fact (cf. Falzea 1965, 432 ff.).[9]

---

the defence of an integrity of the functional environment to the wellbeing of both current collectives and future descendants, and that the explicitly intergenerational viewpoint applied to environmental protection confirms and specifies the anthropocentric vision practised by the legal system. The environment relevant to the legal discipline is not reduced to the environment enjoyed by individuals or by their current communities: the system extends its interest to the human environment as a species" (Cafagno 2007).

[7] The principle of sustainable development, from this perspective, is the main dynamic mechanism of regenerative and adaptive transformation of the legal system: cf. Fracchia (2010), to which reference should be made for complete bibliographical indications.

[8] Legal dynamism which, however, is accompanied by material dynamism. Material reality is changeable, it provides (to the law) ever-new unregulated material, in connection with and as a consequence of those facts that the law evaluates as presuppositions of the regulation.

[9] The relationship of shared evolution between society and the environment leads us to look at the "legal discipline as a process of flexible searching for adaptive paths, the outcome of which depends on the quality of the learning mechanisms that guide it. If contemplated through the lens of the principle of sustainable development, environmental law is a candidate for the role of 'interface' between society and nature, if we like, an artefact which, monitoring and recording ecosystem changes, retroacts on human behaviour in order to promote a permanent process of adjustment of 'historical times to biological times', necessary to safeguard our opportunities for survival, as a species" (Cafagno 2007, 65).

The two profiles into which the analysis of the problem of environmental protection from a legal point of view has been separated are therefore closely intertwined: the means of regulating environmental interests (the response of the law) in fact presupposes the way in which the environment is considered by positive law.[10]

It is helpful to consider briefly the two aspects mentioned: how (and with which characteristics) is the environment relevant for the law? What are the most appropriate legal instruments for the protection of environmental interests?

Environmental positive law would seem not to wish to offer univocal answers to these questions; it provides the interpreter, in fact, with an alluvial, fragmentary, unstable law, one in continuous transformation and flux, within which coexist provisions issued in times that are now distant, most often than not inspired by different underlying philosophies and, often, now outdated; norms of varying natures and levels (international, national, regional, local, of soft law), general and sectoral norms, which often appear to be neither related nor coordinated within a unitary and coherent design.

Nevertheless, it can be stated that the evolution of environmental law has progressively refined the way in which the law relates to the subject of the "environment" and environmental problems.

In fact, from the many and varied positive disciplines there emerge various common traits both with regard to the first profile mentioned above, relating to the way in which the environment is considered by the law, as well as with regard to the formal, substantive, and procedural characteristics of the legal regulation of environmental problems.[11]

2 The Formation of the Legal Notion of the Environment

A brief *excursus* on the slow and gradual process of the formation of environmental law in the Italian legal system[12] might help shed light on some of the pieces that will be useful for an up-to-date understanding of the issue.

---

[10] "The first question with which the interpreter must measure themselves, engaged in establishing whether the environment enjoys an autonomous identity in positive law or if, on the contrary, it is dispersed in a mass of different things, consists in ascertaining whether the rules 'describe' it as a unit that actually transcends the idea of a simple collection of parts; the second problem is to find appropriate 'qualification' criteria. In short, it is first necessary to find a description, then to qualify" [ibid].

[11] First of all, we refer to those general canons which the European Treaty and national law have set up as a common platform of environmental policies, regardless of disciplinary segmentations, such as, for example, in addition to the principle of sustainable development, already referred to above, the principles of precaution, prevention, participation, subsidiarity, and information, which need to be applied within the various areas, taking into account the specificities of the case.

[12] The reconstruction of this path is contained in almost all the environmental law treaties. Here, the setting of Pericu is taken up again (1987, 187 ff.); cf. also Scoca 1979.

It is known that the need to protect the environment began to be felt, on the basis of the first international documents, in the 1960s.

In Italy, the first protection measures were, on one side, from the judiciary, especially the criminal courts and, on the other, a number of bodies of the public administration.

In this first phase, in the absence of laws that directly considered the environment, the legal field employed an expedient: it referred to old rules, enacted to protect entirely different public interests, extending the scope of their application in an interpretative way in order to find in them a new aim of environmental protection. In this way, an initial response to the environmental problem was provided: an exclusively repressive protection, of an episodic and indirect nature, because it was satisfied through the protection of interests other than those of the environment.

Initially, the legislative level adapted very slowly to an awareness of environmental issues: between the 1970s and 1980s, regulatory output was in fact insubstantial and fragmentary.

The first organic intervention on the subject appears with Law no. 319/1979 on water protection (the Merli law), which introduced a series of restrictions on liquid discharges and planning tools for the use and protection of bodies of water.

The most important legislative developments, in this first phase, mainly concern the organisational level: the establishment of the Ministry of Cultural and Environmental Heritage (1975); the conferral to the regions (in 1977) of various functions pertaining to the "protection of the environment"; the establishment of the Ministry of the Environment (Law no. 349/1986); the provision of multiple measures for the connection and coordination of various sectoral interventions at an organisational level; the provision of tools such as Environmental Impact Assessments (EIA), Strategic Environmental Assessments (SEA) and Integrated Environmental Authorisations (IEA), and the responsibility for environmental damage, originally regulated in Article 18 of Law no. 349/1986 (repealed by Legislative Decree no. 152/2006—the so-called Environmental Code—which, among other things, also established new regulations on environmental crime, regarding which see below).

These are all indicators of a tendency on the part of the legal system towards an overall and unitary consideration of the environmental theme.

On the basis of these evolutionary phenomena, the theme caught the attention of doctrine starting in the 1970s, and since then it has studied the possibility of drawing from positive law, in an interpretative way, elements that might be useful for the construction of a unitary legal notion of the environment.

This is a debate, in some ways now outdated, which can be summarised, broadly, by referring to two opposing orientations.

On the one hand, the thesis according to which the legal system, given the state of the legislation (this thesis was elaborated in the early 1970s) attributes to the notion of environment heterogeneous meanings within different disciplinary fields.

The better-known thesis[13] identifies three different notions of the environment: the environment to which the norms regarding the protection of nature and the landscape refer; the environment considered by the norms of defence against pollution; the environment that is relevant in the field of urban planning legislation.[14] According to this approach, positive law does not assume a unitary legal notion of the environment; the "environment" is an expression of synthesis, which refers to specific protections for specific (and different) "environmental assets" (considered separately).[15]

The opposing orientation declared itself in favour of a unitary notion of the environment, but did not reach a definition that was (uniform and) shared of the same. However, it expressed an emerging tendency to systematically look again at the environmental problem and its regulation.

In this context, there is an extremely heterogeneous range of opinions and definitions (the environment as an intangible asset, as a public good (Maddalena 1987, 445 ff.; Maddalena 2010) or public interest of the national community (Postiglione 1985, 32 ff.), as a Constitutional value, as an object of subjective law[16]) which, on closer inspection, are united by a single element: the idea (or perhaps the belief) that environmental regulation had (or could have) a unitary object.

The major criticism that was made of this range of theses is that they forced and pigeonholed the environment into pre-existing categories to justify the applicability of a discipline that appears (each time) to be the one that is adequate (or favoured) and which, in some cases, leads to consequences that are completely contrary to the aim of adequately protecting the environment.[17]

In short, according to the first position, with respect to objective law, it does not consider the environment, but individual components thereof; it is taken into account by specific special norms. The environment, understood as a whole, would instead be considered as the ultimate result (or value) achievable through the practical coordination of these different disciplines.

According to the second position, instead, the environment in and of itself, would be considered as a whole. However, the older theses referring to this orientation

[13] This refers to the thesis of Massimo Severo Giannini (1973, 15 ff.).

[14] Over time the three areas identified by Giannini have gradually accentuated their specialties: in particular, the second is now identified as the core of environmental law although there remains a continuous interaction with and interference from the other two areas: the law of landscape and cultural heritage and territorial government (town planning and construction).

[15] Not dissimilar conclusions are reached by those who identify two (rather than three) profiles of legal importance of the environment (Capaccioli and Dal Piaz 1980, 258), relying on Articles 9 and 32 of the Italian Constitution.

[16] For a summary of the variety of positions that share the point of view of subjective law, cf. above all Dell'Anno (2003, 34 ff.), Dell'Anno (2004).

[17] This refers, in particular, to the proposal to configure the environment (as a unitary asset) subject to subjective law, which is the least suitable situation, both as an individual and a "collective" right, for the adequate protection of the environment. With respect to this problem, the powers, functions, and responsibilities of public bodies, on the one hand, and individual duties and responsibilities, on the other, are of greater importance. Cf. Scoca (1993) and Pericu (1987).

offer inadequate definitions both for the reasons mentioned above as well as for the absence of a mature systemic approach to the theme of the environment.

The evolution since the 1980s and 1990s has seen the multiplication of sectoral regulatory interventions, in which individual environmental assets or the environment, considered in the totality and complexity of its aspects, drawn each time from the norms, have been assumed as objects of relevance and regulation: suffice to mention the law on waste (no. 915/1982); the law on the protection of the sea (no. 979/1982); the Galasso law on the protection of the landscape (no. 431/1985); the framework law on protected areas (no. 349/1991).

The most recent developments in environmental law are inspired, on the one hand, by the reform (2001) of the Fifth Title of the Second Part of the Constitution: the new formulation of Article 117, para. 2 lett. s) now explicitly refers to "the protection of the environment, the ecosystem and cultural heritage"[18]; on the other hand, they are mainly focused on the Environmental Code (Legislative Decree 152/2006 and subsequent amendments): a normative text consisting of over 300 provisions and dozens of appendices, which has been subject to numerous modifications, adaptations, and reforms and which, before the aforementioned interventions to improve it, was, more than a true "code", a jumble of "rules without

[18] The Constitution, before the reform brought about by Constitutional Law no. 3/2001, did not contain any explicit reference to the environment, so much so that the search for a Constitutional support for the functions of environmental protection and the development of criteria for the distribution of the linked competences between the state and the regions involved a complex reconstructive work for doctrine and case law. The Constitutional links were identified in Article 9 of the Constitution, seeing the environment interpretively in the notion of "landscape" mentioned therein (in this sense Predieri who defined the landscape as "the form of the country, created by the conscious and systemic action of the human community [...]" including "every natural pre-existence" and "every human intervention" (Predieri 1969, 381 ff, 387); or in Article 32, which refers to the "right to health", interpreted broadly as including the so-called right to a "salubrious environment" (cf. Giampietro 1980). For a reasoned review, cf. Caravita (1999, 175 ff.). Article 117, letter s), in the text updated by the 2001 reform law, refers today to the subject of the "protection of the environment, the ecosystem and cultural heritage", attributing it to the exclusive legislative power of the state. In the Italian Constitution, no other principle relating to the environment is explicitly stated, nor is there any express reference to the rights and duties of individuals in relation to the environment (though this can be inferred from the duty of solidarity enshrined in Article 2 [see Fracchia 2002]); in an indirect way, the duties of the public institutions can be detected, insofar as, if the disciplining of environmental protection is a state competence, there will be a duty of the state to protect the environment, in the same way as its enhancement will be a duty of the state and the regions. The absence of any formulation of principle on the environment is also accompanied by the absence of references to the principle of sustainable development or the rights of future generations, which are found in other constitutions, such as the French one, after the 2004 amendment, and the German one which, from 1994, in Article 20A, prescribes that the state assumes the protection of natural conditions of life also with regard to future generations. Not surprisingly, despite the Constitutional reform having just been carried out, a proposal was made in 2004 to amend Article 9 of the Constitution, in order to introduce a reference to the environment [cf. on this subject De Leonardis 2004]. The text included the addition of a third paragraph, according to which the Republic "protects the environment and ecosystems, also in the interests of future generations. It protects biodiversity and promotes respect for animals" (draft Constitutional law no. 4307). In this way, the protection of the environment, and this part of the abovementioned Article 117, para. 2, would have been placed among the fundamental principles.

principles", which could not in any way be considered a unitary and coherent discipline of the subject (Fonderico 2006, 632; Celotto 2009; Fracchia 2010, 18).

Furthermore, the evolution of environmental law is not restricted to national law, but also involves international and European Union law: from these disciplinary fields too it is possible to draw important elements for the development of an approach that is more appropriate to the problem of environmental protection from a legal point of view.

3 Environmental Resources and Ecosystem Services

A useful starting point for the specification of the elements that make up the environment, according to law, is the European directive on ecological damage.[19]

This text defines "environmental damage", in general terms, as deterioration caused to natural resources or natural services.[20]

The repair to this damage is mirrored as a restoration of the compromised resources or services or others that are functionally equivalent.[21]

The breakdown of the object that can be damaged into two classes of factors—resources and services—provides a first simple classification of constitutive categories.

Paragraph 12 of Article 2 of the Directive uses the term "natural resource" to designate protected species and natural habitats, water, and land.

This is a partial list, the incompleteness of which, as evidenced by the recitals to the articles, depends on the declared will of the legislation to limit its scope of application, avoiding dealing with behaviours already regulated by special rules or phenomena of environmental deterioration incompatible with the dynamics and properties of the compensatory institutions.[22]

Once it has been established, however, that the category of "resources" contributes to constituting the legal notion of the environment, I believe it is legitimate, for the more general reconstructive ends that are pressing here, regardless of the restrictive

---

[19] Directive of the European Parliament and Council 2004/35/CE of 21 April 2004. The text now takes up and updates the reflection contained in Cafagno (2007), to which we refer, also for a more detailed bibliographic review.

[20] More precisely, Article 2 para. 2 defines damage as "a measurable adverse change in a natural resource or a measurable impairment of a natural resource service, which may occur directly or indirectly".

[21] Para. 11 of Article 2 symmetrically defines "remedial measures" as "any action, or combination of actions, including mitigating or interim measures to restore, rehabilitate or replace damaged natural resources and/or impaired services, or to provide an equivalent alternative to those resources or services". A more detailed breakdown of the remedial measures, faithful to the basic distinction between resources and services, is contained in Annex II of the Directive.

[22] For example, episodes of "general and widespread pollution, in cases where it is impossible to link negative environmental effects to acts or omissions of certain individual subjects"; thus considering no. 13.

view imposed by the contingent choices on the compensatory regime, to extend the attention to the typological class as a whole.

An organic review of the elements included in the class is offered, one of many official texts, by the Communication of the European Commission entitled *Towards a thematic strategy on the sustainable use of natural resources.*[23]

The document, first of all, enumerates the "raw materials such as minerals (including fossil energy carriers and metal ores) and biomass".

The second group included in the category consists of "environmental media, such as air, water and soil", i.e. the physical *media* within which life takes place: atmosphere, geosphere, and hydrosphere.

The third consists of "flow resources, such as wind, geothermal, tidal and solar energy".

Finally, "space" follows, which is required "to produce or maintain all the above-mentioned resources".

In short, raw materials, biomass and biological organisms, air, water and soil, flow resources, space and land, are the material elements included in the legal catalogue of natural resources.

The ambivalent predicate "natural" that qualifies the noun "resources" serves to circumscribe a range of assets that nature makes available to human beings, that is, that are not a product of anthropic action, but its free support.

The reconstructive effort of the legal concept of environment must now continue with the step from the preliminary inventory of resources—necessary, but not sufficient—to the category of "natural services", a second constitutive element of the damage, the study of which brings the systems theory back to the centre of the discourse.

The explicit reference to services, within the norms, frees the legal configuration of the environment from the sterile schema of a list, enhancing a functionality that places interdependencies at the heart of the discourse.

In fact, it is known that, unlike the units of a whole, the parts of a system are not considered as mere addenda, since they import their reciprocal relationships[24]; by virtue of the organisational connections, disparate bodies assume a unitary identity, which makes it possible to discern new and autonomous properties, which cannot be deduced from the separate observation of the single components (with obvious reference to Von Bertalanffy (2015); cf. also Capra (1983), Holland (1992, 1996, 2000, 2013, 2014) Simon, 1962).

In short, these norms testify that the environment is not only recognised by the law as a "container" of resources, but also because it provides for their continuous regeneration through a complex network of interlaced processes and, more generally, for the conservation of that interval of physical conditions within which humans, among other living beings, can exist.[25]

---

[23] Communication from the Commission, COM (2003) 572 final, cit.; cf. also the following Communication COM (2005) 670.

[24] Obviously referring to L. Von Bertalanffy (2015).

[25] From this perspective, the European Environment Agency (1999), Costanza et al. (2001), Berkes

The ecosystem functions are perpetuated autonomously, through the continuous reciprocal adaptation of the abiotic and biotic constituents.

The classification of services as systemic, the fruit of spontaneous self-organisation, subject to common use, sets the scene for the clarification of the logic of public intervention and facilitates the understanding of the specific elements of environmental administrative law.

This boundary between normative logics reflects a distinction that is one of the cornerstones of the so-called ecological economy; the line of thought invites us to consider environmental resources according to two complementary points of view: as stocks, that is, as provisions that generate flows, subject to collection, and as systemic components, responsible for the perpetuation of vital services (Georgescu Roegen 1973; Daly and Farley 2004, 106 ff.; Costanza et al. 1997, 53 ff.; Costanza & Farley 2010; Daily 1997; De Groot et al., 2002; Duraiappah 2014; Fari, 2013; Gunderson & Pritchard 2002; Gustafsson 1998; Hester & Harrison 2010).

4 The Environment as a Shared System

Services such as climate regulation, carbon fixation, the nitrogen or phosphorus cycle, photosynthesis or biomass production, biodiversity conservation, precipitation regulation, sea level control, maintenance of viable levels of oxygen, the transmission of values inherent to the aesthetic perception, or form of the land are examples of utilities that lack the precondition of excludability.[26]

Other constitutive elements of the environment, among those enumerated in the previous paragraphs, would seem—on the other hand—to lend themselves, at first glance, to an antagonistic and exclusive consumption.

In fact, the legal system confers individual property rights or exclusive usage rights on assets such as parcels of land, livestock, mineral deposits and crops, which are abstractly susceptible to exchange under the market and common law regime.

However, even resources apparently subject to division and differentiated usage can assume the relevance of commons—or at least complementary elements of partially indivisible assets—as soon as we start to consider them, according to the logic just mentioned above, from the systemic perspective (Berge 2003a, b).

The portions of timber taken from a wooded area, the food or medicinal substances derived from animal or plant organisms, the quantities of raw materials or minerals extracted from deposits to be refined and processed, the volume of water

et al. (2002), Gunderson and Holling (2002), Allen and Starr (1982), Bologna (2005, passim and 117 ff.), Lee (1993a, b), Levin (1999), Marten (2001).

[26] On the general topic of public assets, in the economic sense, among the numerous writings, cf. Stiglitz (2000); in particular on the commons—as assets materially distinguished by the combination of the two properties of "non-excludability" and "rivalry in consumption"—the reference to Ostrom (1990, 1996) cannot be avoided; cf. also, among many, Barnes (2006), Berge (2003a, b), Daly and Farley (2004). In Italian doctrine, among many, Boscolo (2012), Bravo (2001, 487 ff.), Cafagno (2007), Maddalena (2013, 91 ff.), Marella (2012).

drawn from a source to be bottled and sold, or an area destined to be a receptacle for waste from industrial plants are all banal examples of flow units, directly or indirectly incorporated into goods, the object of common exchange.

As we have seen, the quantities of resources cannot however be reduced to simple inventories since they deploy the contextual action of components of the environmental system.[27]

The exploitation and commercialisation of the flow units has therefore to be combined with safeguarding the functional and structural integrity of the system that feeds the flow and, more generally, ensures the relative services.

The components of the biosphere—or the ecosystems that are housed in it—are revealed in this light as parts of functionally unitary objects, responsible for services in support of life, which present with profiles of indivisibility.

It is true that, for the purposes of usage and construction, a resource such as the land is usually divided and subdivided into parts, made the object of ownership, and as such exploited.

But when the same resource is considered for its landscape functions or ability to influence the dynamics of animal or plant populations, the hydro-geological structure or the consequences of the relative transformations, the boundaries traced by criteria of ownership or rules of use, of a civil or planning nature, are inevitably crossed and the land as a whole returns to being important as a system or part of a wider system, in the reproduction of the services of which its various parts (separate for other purposes) participate organically (Berge 2003a).

It is no coincidence that the aforementioned directive on the illicit states that "environmental protection is [...] a diffuse interest on behalf of which individuals will not always act or will not be in a position to act"[28] and it excludes cases of personal injury, damage to private property, as well as compensation for traditional damage under civil liability rules.[29]

Ultimately, the conduct regulated by environmental law would seem to all have in common the aptitude to affect—favourably or harmfully—a system or its components with shared access, at the origin of the propagation of natural services, unable to be enjoyed selectively and in a discriminatory way.

Paradoxically, the more the flow units fed by the stocks of natural resources are freed from the attribution of non-excludability, that is, they lend themselves to being removed and exploited, to be incorporated into the so-called private assets, perhaps circulating on the market, the more precarious and problematic the safeguarding of the system becomes, as a functionally organic structure (Ostrom 1996).

In other words, a latent conflict emerges in all its criticality between the incentives triggered by the calculation of the utilities achievable through the use of the environment and the limits imposed by the awareness of a value of existence

---

[27] Ostrom (1996, 9 ff.) clarifies: "an irrigation system, a pasture, a mainframe computer or a bridge are all examples of systems of common, manufactured or natural resources. Water, fodder, central computing units and units crossing the bridge are all examples of unitary flows of extractable resources".

[28] Recital no. 25.

[29] Recital nos. 11 and 14.

(Leakey & Lewin, 1995; Millennium Ecosystem Assessment 2003, 128 ff., spec. 132–133; Weisbrod 1964, 471 ff.; Pearce and Turner 1990).

Natural resources and services have a value both as a result of appropriation and current use, insofar as they maintain the role played in the ecosystem organisation, benefiting not only those who might consume them in the present, but also those who will come after (here the so-called legacy value appears) (Millennium Ecosystem Assessment 2003, 132 ff.; Arrow and Fisher 1974, ff.; Pearce and Turner 1990).

The fact that the legal system grants decisive importance to the values of existence, functional to the well-being of the communities today and those of the future, is once again testified to by the centrality of the principle of sustainable development.

5 A Summarising Formula: Environmental Objects Are Considered by Law as Physical Commons with a Legacy Value

A desire for clarity leads us to summarise the steps of the reasoning.

The discipline of ecological damage invites us to consider the environment as a system that combines organic and inorganic elements (the "natural resources") from the complex interaction of which emerges a spontaneous flow of "services", diffuse and destined for collective use.

This is a perspective shared by the main institutes of environmental law with a character that is general and transverse.

Simply by way of example, without any pretence of completeness, Article 5 para. 1 letter c) of Legislative Decree no. 152 of 2006, regarding EIAs and SEAs, in defining the notion of environmental impact, describes the environment as a "system of relations", in line with the relevant European directives.

Article 2 of Directive 2003/4/EC of the European Parliament and Council of 28 January 2003 describes as "environmental" information relating to air, atmosphere, water, soil, land, landscape, natural sites, coastal and marine areas, biological diversity and, alongside these, the "interactions between these elements".

An equivalent formulation is contained in Article 2 of Legislative Decree no. 195 of 19 August 2005, transposing the directive.

The European Convention signed in Florence on 20 October 2000 defines the landscape as "part of the land, as perceived by the populations, the character of which derives from the action of natural and/or human factors and their *interrelations*" (Herrero De La Fuente 2001, 893 ff.).

If observed as systems, from the interconnection of which emerge vital services, the biosphere and linked ecosystems become individually inappropriable functional units, the components of which, however, are subject to antagonistic and differentiated uses.

The existence, in the environment and in many of its structural elements, of the typical features of the so-called commons (high costs of exclusion accompanied by rivalry in consumption) creates conditions favourable to the compromising of the integrity of the system.

The defining element of the environmental dilemma is represented by the fact that the danger fomented by shared use cannot be avoided by merely conservative policies because the preparatory use for development (which is different from growth) remains indispensable.

The discipline of the environment is developed around the problematic nucleus identified in the provocative article by Hardin (1968) as the "tragedy of the commons".[30]

It is a normative the essential purpose of which is identified in the balance between the need to use environmental resources and the need to preserve the performance capacity of the system that they contribute to form, to which the system recognises, alongside a usage value, a value of existence and legacy, for the benefit of present and future generations.[31]

6 Systemic Vision and Legal Principles to Protect the Environment

Within the coordinates set out above, the principle of sustainable development invites us to look at the environment and human society as two subsystems, which interact in the biosphere, a broader system that brings them together.

The subsystems continuously exchange energy, information and matter and, consequently, undergo constant change (Costanza et al. 2002, 409–420; Prigogine and Stengers 1989, *passim* and p. 44. ff.; 1999, *passim* and 144 f.; Prigogine 1988, 52 ff.; Tiezzi 1988, 442 ff.).

The critical node around which the sustainability imperative orbits can therefore be represented as a problem of mismatch, in time or space, between the scale of human actions, processes and responsibilities, and the scale of the processes and dynamics of the ecosystem.[32]

It is clear that the task of guiding and combining collective decisions and individual choices so as to promote the harmonisation of its complex systems, that is, of socio-economic and ecosystem dynamics, presents prohibitive difficulties; reasonably, a field like this is not "governed" in the oligarchic and deterministic

[30] "It is easy to write laws that establish bans (even if it is not easy to enforce them); but how can we write laws that sanction temperance?" (Hardin 1968, 1243 ff.).

[31] For a more in-depth examination of the thesis that the legal notion of the environment, like a systemic vision, aggregates objects that are considered as commons with a legacy value, cf. Cafagno (2007, spec. 146 ff.).

[32] Above all, cf. Lee (1993a, 560 ff.; b); also recommended is a reading of Costanza et al. (2001, 8 ff.), Folke et al. (2007), Young (2002, *passim*), Cumming et al. (2006, 14), Wilson (2006). As for the typically hierarchical structure of complex systems, it is sufficient to refer to Ahl and Allen (1996), Simon (1988, 208 ff.).

sense of the word (Allen et al. 2003). If anything, the institutional effort to provide a virtuous trajectory to the evolution of changes in need of harmony is urgent (Lee 1993b).

It has been rightly noted that in this matter, marked by uncertainty and complexity, it is often necessary to think of rules and legal choices as working hypotheses, rather than solutions. An evolutionary and adaptive approach requires that the hypotheses are methodically verified and corrected, through apparatus and procedures capable of accumulating and processing the information drawn from the observation of the results achieved, i.e. suitable to trigger and stimulate learning dynamics.

In short, complexity poses questions for the jurist that do not admit responses, but processes of responses (cf. Weizsacker and Weizsacker 1988, 126 ff.).[33]

The link between society and the environment—in time and space, coordinating the sphere of knowledge, decisions, and incentives—translates in synthesis into the refinement of adaptation mechanisms (cf. Rammel and Van den Bergh 2003, 121 ff.; Ramos-Martin 2003, 387 ff.).

The propulsive agent of the so-called adaptive management–a method of managing environmental problems that represents one of the pillars of the ecosystem approach[34] and which an increasing number of authors and scholars consider the most appropriate—is learning.

It is with this philosophy that the legal principles that govern environmental matters ought to be studied.[35]

The critical knot of uncertainty calls into question two symmetrical needs: caution and learning.

Both require adequate information processing (Licata 2013).

Only a timely recognition of the feedback, on the interweaving of which resilience depends (Holling 1973, 23 ff.; 1988, 115 f.], and the performance capacity of the environment allows us to act with caution and learn at the same time.

Secondly, given that the functions and structure of landscapes and ecosystems arise from the reciprocal action of entities and processes that operate on multiple scales of time and space, the design of equipment, skills and administrative procedures should coherently reproduce variety and redundance.

In other words, the legal system should also operate on multiple scales (so as to be able to duplicate the relevant environmental and social feedback at each scale), through coordinated and interlaced responses (so that the interdependencies between the different organisational levels of the ecosystem and human communities are not lost) (Ostrom 1998, 149 ff.; 2005; Dietz et al. 2003, 1907 ff.; Ruhl et al. 2007).

[33] For an analysis of the problem of complexity management in administrative law cf. D'Orsogna (2002, 2003, 2005).

[34] An effective synthesis of the conceptual basis on which the so-called ecosystem approach is based is offered by Decision VII/11 of the 7th Conference of the Parties to the Convention on Biodiversity, held in February 2004 in Kuala Lumpur (http://www.biodiv.org); in doctrine, among many, Christensen et al. (1996, 665 ff.).

[35] For the illustration of which Crosetti et al. (2002).

In general, systems theory states that a proportionate degree of complexity is required to respond to complexity; Ashby's law of minimum variety teaches us that a control system cannot possess less variety and less versatility than the system it seeks to control (Ross Ashby 1956, 202 ff.).[36]

A strictly centralised institutional apparatus, dominated by a logic of command and control, would lack the indispensable presuppositions for the task of integration, even if meticulously constructed (Bar-Yam 2004b, 37 ff.).

On this basis, it is not difficult to grasp the advantages of a legal design based on the logic of autonomy and decentralisation, allied with flexible collaborative and participatory mechanisms, with the support of a versatile and creative recourse for the mixture of authority and market.

On the other hand, since the environment has the nature of a shared-access system, it is essential that—in line with what is generally the case with the commons—legal and regulatory tools are able to smooth and lessen conflict, gaining credit and trust, facilitating the formation of shared choices and the ongoing observance of cooperative and realistic decisions, pragmatically based on dialogue between the social partners (Ostrom 1990, 1996).

Reflecting the first node are the legal principles—of prevention, precaution, correction of damage at the source and information—which, in imposing a prudent management of risks and dangers, pursue the goal of integration into the temporal dimension.

The principles of subsidiarity, co-responsibility, loyal cooperation, participation and, once again, information are easily linked to the second and third nodes, the main purpose of which is to harmonise the social and natural system on the organisational and territorial level.

Going from this general framework to a closer examination of current law, the following paragraphs will seek to focus attention on the emblematic and exemplary subject of climate change.

7 Environmental Law Tested by Complexity: Climate Change

If we want to consider the characteristics of the environmental discipline, which has now arrived at a very significant breadth and complexity rarely found in other sectors, it must be agreed that the traditional approach, at least in the first phase of development of the sector, is marked by a fairly frequent recourse to command and control, as it is known.

In substance—and this is particularly evident in the regulation aimed at protecting the environment from pollution: air, water, waste, and so on—the law provides for the setting of standards, limits, or general prohibitions; it then subordinates the

[36] For further information and applications cf. Bar-Yam (2004a, b, 37 ff.).

exercise of private activities to a permissive provision (often in agreement with a plan) and assigns to the public authority duties of control (on compliance with limits and authorisations) and the power to sanction.

The command and control model works very well in certain cases, but displays a number of obvious limitations. Firstly, it is very rigid and requires a generalised application (it should be intolerable that in some regions of the country the limits are respected and in others not; all of which also implies the presence of a competent and efficient administrative organisation); secondly, the model is difficult to implement, insofar as it is hindered by the problem of information asymmetries that often affect the public entity called on to set the optimal level of standards; the modification and adaptation of the system as not immediate either, since they imply that the issue is brought to the attention of the political sphere and the adoption of decisions which, as a rule, are the result of complicated negotiations and difficult compromises; finally, command and control does not stimulate the individual responses of the more virtuous operators on the market.

Precisely in order to overcome these difficulties, in the literature (and, within certain limits, in the legal system) different environmental protection models have been developed, which, with a view to greater flexibility, and in order to encourage virtuous behaviour, enhance the market and its dynamics. There are many variations: from taxes to subsidies, from green public procurement to certification systems, all the way to the creation of artificial markets.

It is necessary to ask whether, in weighing up the various models, the legislator and the public decision-maker have given proper consideration to (and had an awareness of) the complexity of the framework of reference, or if they have organised an excessively rigid and therefore inadequate response, because it was calibrated on a single scale of intervention.

In order to test the "degree of usage" of a correct approach to environmental complexity, it would appear to be useful to turn the attention to a concrete question, a very complex and global one, namely climate change, now very much at the top of the political agenda.

On the other hand, looking at the results of the 5th IPCC Assessment Report,[37] released in the period 2013–2014, it is difficult not to have serious concerns about the future of our planet.

The Report, which is very complex, articulated and not always easy to read, emphasises that the warming of the climatic system is unequivocal; there is a warming of 0.85° in the period 1880–2012, a rise in sea level of 0.19 m, a carbon dioxide concentration that has increased by 40% from the pre-industrial age (among other things, compared to the past, the increase in the contribution of Asia is frightening), underlining how it is extremely probable that human influence has been the dominant cause of the warming observed since the twentieth century, confirming the absolute need to consider the complexity of the market and its dynamics.

One of the most frightening data is the probability that the change in global surface temperature by the end of the twenty-first century will exceed 1.5 °C compared

[37] Consultable at the following address: https://www.ipcc.ch/report/ar5/.

to the period 1850–1990; according to the worst-case scenarios (in other words, in the event that emissions are not reduced), however, overheating could be between 2.6 and 4.8 °C by 2100.

The tolerability limit for the increase in the temperature of the planet is cautiously set at 2 °C: to maintain it within those limits considerable effort is necessary (the key year to reverse the trend and reduce overall gas emissions is set at 2030).

Further concerns, in Italy, might arise when looking at the First and Second Reports on the State of Natural Capital in Italy, published in 2017 and 2018 pursuant to Law no. 221/2015 by the Committee for Natural Capital.

It is, in any case, interesting to look at the evolution of the legal response, since the climate change sector is one of those in which we have tried to learn from mistakes and propose innovative lines of action.

8 Climate Change and the Legal Response

At global level, the response to the problem of climate change occurred initially through the Kyoto Protocol, aimed at reducing polluting emissions in the atmosphere. These are, in particular, greenhouse gases, that is, those gases (including carbon dioxide), mainly deriving from the combustion of fossil fuels, which prevent the irradiation of energy in the atmosphere, causing an overheating of the planet.

The Protocol was the legal basis for a complex political-institutional undertaking, initiated with the multilateral negotiation conducted within the *Intergovernmental Negotiating Committee for a Framework Convention on Climate Change* created by the United Nations General Assembly. This the concrete legal instrument linked to the United Nations Framework Convention on Climate Change adopted in New York in 1992 and signed in Kyoto during the Third Plenary Session of the Conference of the Parties in 1997.

The Protocol, although not ratified by a number of important countries (the USA first and foremost, which considered the commitments for industrialised countries to be too burdensome, whereas other major polluters—China and India—were excluded from Annex I, related to the industrialised countries), still entered into force with its ratification by Russia in 2005. The European Union adhered to the commitments (the Kyoto Protocol of 1997, more specifically, was ratified by the individual countries—in Italy with Law 120/2002—and the Union).

Among the instruments envisaged by the protocol, Emission Trading, based on the principle of cap and trade, is worthy of special mention: the setting of a maximum limit on emissions and market dynamics.

The contracting parties undertook to reduce emissions by 5% between 2008 and 2012 compared to 1990 levels. The objective was then divided into the various regions and, in the different sectors involved, between the operators.

Activities involving greenhouse gas emissions falling within the scope of the Protocol cannot be carried out without specific authorisation, with sanctions for cases of violation of the obligation, which can go as far as authorisation being revoked.

Up to this point, there is no market intervention.

It "enters the stage" with reference to business strategies and should favour the more efficient companies, taking into account that a quota deficit is sanctioned, while a quota surplus can be traded on the market.

In order to meet the obligations (and, that is, not to pollute to a greater extent than the quotas that have been distributed to them), companies, in fact, can decide to reduce emissions, for example, by investing in innovation, or, in the case of emissions higher than the permitted limits, by buying quotas from other operators (evidently on the basis of a cost-benefit calculation). Instead, where there are particularly "virtuous" subjects (i.e., who are able to emit a quantity of gas that is lower than what is permitted), they may sell the excess quotas or capitalise them, setting them aside for the following years.

The market is therefore the place where, in view of the authoritative determination of the extent of tolerable pollution, quotas are exchanged and the price is set. In order to verify compliance with the limits, companies must finally "return" a number of quotas equal to those obtained each year, otherwise administrative sanctions will be imposed.

In this chapter, we are not interested in analysing in greater detail the functioning of the Protocol, nor referring to the imposing pile of literature that has been formed in this regard. It is sufficient to note that, within the framework of a very complex normative architecture, there is a mix of authoritative powers and market instruments.

Thus, we can return to the initial question: did the functioning of the Kyoto Protocol, the implementation of which resulted in significant institutional and regulatory efforts and the assumption of onerous obligations, lead to a real improvement in the conditions of the environment? If this did not happen, did the failure depend on a wrong approach to the problem of climate change?

9 The Limits of the Approach to the Problem of Climate Change

There are several criticisms that can be made of the approach used to reduce the emission of greenhouse gases.

Firstly, the mechanism of the Protocol makes sense in cases where pollution is territorially indifferent, precisely because its implementation might result in a global decrease in emissions, but generate a local concentration of pollution.

Furthermore, the value of the quotas is significantly lower than expected, so that companies have been encouraged to buy them on the market rather than make investments. This occurred for many reasons, not least the fact that the initial distribution was probably wrong. The quotas, in an early phase, were in fact assigned free of charge on the basis of the emissions history (the so-called grandfathering), thus excluding market intervention already in this delicate phase, with the result, however, that the most polluting and less virtuous companies ended up getting more quotas. Originally, the allocation was by auction, which implies an effort by the most polluting industries to acquire permits and generates the availability of resources for states as a result of the auction. On the other hand, too many quotas have probably been allocated, decreasing their value, thus weakening the intensity of the operators' efforts to respect the tolerated pollution limit (since it was more convenient to purchase quotas than to invest).

These limits (in particular, the second, because the former has to do with the fluidity of the decisions of the players), while undoubtedly very important, are not yet directly linked to the deficit of an adequate approach to the complex nature of the problem we hypothesised had characterised the response of the law.

On the other hand, the last characteristic emphasised emerges clearly when other aspects are considered.

A global tool for combating climate change would require the unanimous adhesion of all polluters. This did not happen, generating the frustration (and doubts) of the states strongly committed to emission reduction efforts, which risk being thwarted by the growing pollution of other countries, the developing ones which are not subject to the limits of the Protocol. The rigidity of the legal mechanism, that is, while aspiring to foreshadow a global response, led to a sectoral approach (in terms of states involved and activities concerned) and failed to coordinate (and be coherent) with the complexity and pervasiveness of the problem.

Moreover, the possibility of generating tolerable pollution has been incorporated into quotas, movable assets traded and managed by financial intermediaries, to which considerable wealth has been transferred. It could be doubted (and indeed it was) whether, at least in Italy, the mechanism generated investments in favour of environmental protection. This perspective has ended up transforming environmental problems into a financial dimension, without being able to generate adequate incentives for investments. However, in hindsight, this reflects the fact that the legal mechanism proved to be too rigid and unable to consider the flexibility of the market. In other words, there is a lack of coordination between the legal dimension and the fluidity of economic dynamics: the market does not respect the "indications" provided by the law and tries to pragmatically exploit the investment opportunities that arise, including and especially those resulting from insufficient regulation, which was not able to "learn" from the object and adapt adequately.

Lastly, very evident limit emerges considering that the regulation of the permit market is based on certain types of production and not on consumption. This regulatory asymmetry (or, better, insufficiency) has induced the players in the market (who, as a consequence of the Kyoto Protocol, would have to bear higher production costs and, therefore, immediately repositioned themselves, with a speed that the law was unable to contain) to invest in states not forced to respect the limits (developing countries). The consequence is that global pollution is not decreasing, while consumers in developed countries continue to request goods that are imported, generating further costs for the environment and damage to local economies.[38]

[38] With reference to China, for example, cf. Weber et al. (2008, 3572–3577), Peters (2008, 13–23).

10 The Most Recent Developments (Outline of the Paris Agreement)

The example of the fight against climate change provides a series of interesting indications.

The rigidity and sectoral nature of the regulatory system (even if inspired by the desire to organise a global response), for example, have shown all their limitations in the face of a vast, complex problem characterised by the presence of two dimensions (environment and market) in constant evolution, to which the law must adapt. The market, in particular, proved to be more than ready to exploit the "flaws" of the legal system and adapt to new developments.

Moreover, the global scale adopted and the top-down approach have not allowed the "capture" of the infinite variety of concrete behaviours which, in the same way as big industrial decisions, generate climate change. The complexity of the problem and of the systems involved (environment and market) has not, therefore, been matched by a similar variety of scales and approaches to govern it. In general, the environmental management method that goes by the name of adaptive management and represents one of the cornerstones of the so-called ecosystem approach is absent (Cafagno 2013).

Law and politics have partly perceived the need to use a different approach and tried to react.

The Kyoto Protocol foresaw the 5-year period 2008–2012 as a temporal horizon of reference to verify the achievement of its binding reduction targets by the states listed in Annex I of the Protocol. Numerous efforts were made to design a path and a shared legal regime related to the post-2012 period (powerful expectations had, for example, fuelled the 2009 Conference of the Parties in Copenhagen, which, on the contrary, produced only a partial agreement of little legal significance).

We thus reach the Paris Climate Agreement,[39] adopted on 12 December 2015 and which entered into force on 4 November 2016 (in reality, it is composed of two documents: the Decision and the Agreement, which formally constitutes an annex to the former and which is the only act that is binding and subject to ratification).

The shared goal was that of containing the increase in the global average temperature within 2 °C, while at the same time making every effort to achieve an even more ambitious result (1.5 °C).

The distinction between countries in different groups and burdened by differentiated duties that had characterised the "model" of the Kyoto Protocol was avoided (even if, in some respects, a different role still remains for various countries): each party, in fact, has to take national contributions, differentiated, but not binding, of mitigation (but the goal of increasing the ability to adapt to climate change and climate resilience is also very important).

There is no shortage of market instruments, only within certain limits similar to those already provided for by the Kyoto Protocol: on the one hand, cooperative

[39] Cf. Nespor (2016, 81 ff.) and Montini (2017, 317 ff.); Rajamani 2016; in general, see The Paris Agreement (2016).

approaches are envisaged, which can be used on a voluntary basis by the contracting parties to achieve their national mitigation objectives (the transactional transfer of emission reduction units through the exchange of Internationally Transferred Mitigation Outcomes, linking more market systems in the name of greater flexibility); on the other hand, under the authority and guidance of the Conference of the Parties, a clean development mechanism is envisaged to promote the logic of sustainable development: the reduction of greenhouse gas emissions obtained in the territory of another contracting party can be accounted for to achieve the objectives set by its own national mitigation contributions.

Jurists and scholars of the environmental problem, theoretically committed to further seeking a scientific and epistemological way of behaving which is appropriate to the complexity of the object, are called on to exercise careful criticism and continuous vigilance on the functioning of the new instruments.

For example, they cannot avoid criticising the fact that many of the provisions of the Agreement do not impose specific and binding obligations of conduct or results. This is particularly evident in relation to the objectives of the states. This is a significant deficiency, also taking into account that the time to effectively intervene to avoid the most worrying scenarios continues to flow dramatically and is very limited, while the Agreement foresees a regime that will substantially enter into force from 2020.

Another weakness is linked to the absence, within an agreement that aims to counter a global problem, of important players, especially following the abandonment of the USA.

Having clarified this, it is undoubtedly difficult now to formulate a more articulated judgment in the absence of certain data on the implementation of the model and before the precise rules are set for its operation. In particular, it will be necessary to verify within which limits the intervention scales will be multiplied and the balance between command and control and market-based tools.

However, we cannot fail to note how the Agreement has tried to take into account some of the needs (closely related to each other) that were previously underlined and, therefore, reflects a greater awareness of the complex nature of the problem: it is sufficient to mention the need to operate a periodic review of the interventions, which will take place every 5 years, and the overcoming of the rigidity of the Kyoto Protocol, also due to the abandonment of a top-down approach (based on the imposition of binding targets on a group of countries).

In other words, a more flexible system has been chosen, characterised by a bottom-up structure: instead of specific and timely commitments, it was decided to provide a generic duty of collaboration and, above all, the previously mentioned system of non-binding mitigation contributions of greenhouse gas emissions, defined unilaterally and voluntarily by the individual contracting parties (NDCs: Nationally Intended Contributions): mechanisms with less binding force, but which leave more scope for the use of a multiscale approach.

References

Ahl, V., & Allen, T. F. H. (1996). *Hierarchy theory: A vision, vocabulary, and epistemology.* New York: Columbia University Press.

Allen, T. F. H., & Starr, T. B. (1982). *Hierarchy-perspectives for ecological complexity.* Chicago: The University of Chicago Press.

Allen, F. H., Tainter, J. A., & Hoekstra, T. W. (2003). *Supply-side sustainability.* New York: Columbia University Press.

Arrow, K. J., & Fisher, A. (1974). Environmental preservation, uncertainty and irreversibility. *Quarterly Journal of Economics, 89*, 312–319.

Barnes, P. (2006). *Capitalism 3.0.* Retrieved from http://capitalism3.com.

Bar-Yam, Y. (2004a). *Making things work. Solving complex problems in a complex world.* Cambridge: NECSI, Knowledge Press.

Bar-Yam, Y. (2004b). Multiscale variety in complex systems. *Complexity, 9*(4), 37–45.

Benvenuti, F. (1982). Coordinamento concetto o coordinamento schema? In G. Amato & G. Marongiu (Eds.), *L'amministrazione della società complessa: In ricordo di Vittorio Bachelet.* Bologna: Il Mulino.

Berge, E. (2003a). Commons: Old and new on environmental goods and services in the theory of commons. In E. Berge & L. Carlssonm (Eds.), *Landscape, law & justice: Proceedings from a workshop on old and new commons.* Oslo: Centre for Advanced Study. Retrieved from http://dlc.dlib.indiana.edu.

Berge, E. (2003b). *Environmental protection in the theory of commons.* Paper presented at the Congress *Transnationalizing the Commons and the Politics of Civil Society*, Chiang Mai, Thailand, 11–14 July. Retrieved from http://dlc.dlib.indiana.edu.

Berkes, F., Colding, J., & Folke, C. (Eds.). (2002). *Navigating social-ecological systems: Building resilience for complexity and change* (1st ed.). Cambridge: Cambridge University Press.

Bertalanffy, L. (2015). General system theory: Foundations, development, applications (rev. ed.). New York: George Braziller.

Bocchi, G., & Ceruti, M. (2004). *Educazione e globalizzazione.* Milano: Raffaello Cortina Editore.

Bologna, G. (2005). *Manuale della sostenibilità. Idee, concetti, nuove discipline capaci di futuro.* Milan: Edizioni Ambiente.

Boscolo, E. (2012). *Le politiche idriche nella stagione della scarsità. La risorsa comune tra demanialità custodiale, pianificazioni e concessioni.* Milan: Giuffré.

Bravo, G. (2001). *Dai pascoli a internet. Un'introduzione alle risorse comuni.* In Stato e Mercato, no. 63.

Bucello, M., & Cafagno, M. (2005). *Inquinamento, in Nuovo Digesto, disc. pubbl.* Turin: Utet.

Cafagno, M. (2007). *Principi e strumenti per la protezione dell'ambiente. Come sistema complesso, adattativo, comune.* Giappichelli: Turin.

Cafagno, M. (2013). Cambiamenti climatici tra strumenti di mercato e potere pubblico. In F. Cartei (Ed.), *Cambiamento climatico e sviluppo sostenibile.* Giappichelli: Turin.

Capaccioli, E., & Dal Piaz, F. (1980). Ambiente (tutela dell'). Parte generale e diritto amministrativo, under *Appendice Noviss. Dig. it.* Turin: Utet.

Capra, F. (1983). *The turning point: Science, society, and the rising culture.* London: Flamingo.

Caravita, B. (1999). *Costituzione, principi costituzionali e tecniche di formazione per la tutela dell'ambiente.* In Grassi, Cecchetti, Andronio [1999].

Caravita, B. (2005). *Diritto dell'ambiente.* Bologna: Il Mulino.

Cecchetti, M. (2000). *Principi costituzionali per la tutela dell'ambiente.* Milan: Giuffrè.

Celotto, A. (2009). *Il codice che non c'è: il diritto ambientale tra codificazione e semplificazione.* Retrieved from www.giustamm.it.

Christensen, N. L., et al. (1996). The report of the Ecological Society of America Committee on the Scientific Basis for Ecosystem Management. *Ecological Applications, 6*, 665–691.

Cioffi, A. (2009). *L'ambiente come materia dello Stato e come interesse pubblico. Riflessioni sulla tutela costituzionale e amministrativa*. In Riv. Giur. Ambiente, no.6.

Civitarese Matteucci, S. (1992). *Ancora sulla nozione giuridica di ambiente e sulla legittimazione delle associazioni ambientalistiche all'impugnazione degli atti amministrativi*. In Riv. Giur. Ambiente, no. 3.

Civitarese Matteucci, S. (2003). *Il paesaggio nel nuovo Titolo V, Parte II, della Costituzione*. In Riv. Giur. Ambiente, no. 2.

Cordini, G. (1994). *La tutela dell'ambiente: il contributo della ricerca multidisciplinare*. Padova: Cedam.

Cordini, G. (2012). Diritto ambientale comparato. In P. Dell'Anno & E. Picozza (Eds.), *Trattato di diritto dell'ambiente* (Vol. I). Padua: Cedam.

Costanza, R., & Farley, J. (2010). Payments for ecosystem services: from local to global. *Ecological Economics, 69*(11), 2060–2068.

Costanza, R., Cumberland, J. H., Daly, H., Goodland, R., & Norgaard, B. (1997). *An introduction to ecological economics hardcover*. London: CRC Press.

Costanza, R., Low, B., Ostrom, E., & Wilson, J. (Eds.). (2001). *Institutions, ecosystems, and sustainability*. Boca Raton: Lewis.

Costanza, R., Limburg, K. E., O'Neill, R. V., & Farber, S. (2002). Complex systems and valuation. *Ecological Economics, 41*(3), 409–420.

Crosetti, A., Ferrara, R., Fracchia, F., & Olivetti Rason, N. (2002). *Diritto dell'ambiente*. Rome: Laterza.

Cugurra, G., Ferrari, E., & Pagliari, G. (Eds.). (2006). *Urbanistica e paesaggio*. Editoriale scientifica: Napoli.

Cumming, G. S., Cumming, D. H. M., & Redman, C. L. (2006). Scale mismatches in social-ecological systems: Causes, consequences, and solutions. *Ecology and Society, 11*(1), 14. Retrieved from www.ecologyandsociety.org/.

D'Amelio, P. (1988) under *Ambiente (diritto amministrativo)*. In *Enciclopedia giuridica Treccani* (1st., Vol. II). Rome: Enc. It.

D'Orsogna, D. (2002). *Conferenza di servizi e amministrazione della complessità*. Turin: Giappichelli.

D'Orsogna, D. (2003). Una terapia sistemico-relazionale per la pubblica amministrazione: l'operazione amministrativa. In M. Immordino & A. Police (Eds.), *Principio di legalità e amministrazione di risultati*. Turin: Giappichelli.

D'Orsogna, D. (2005). *Contributo allo studio dell'operazione amministrativa*. Naples: Editoriale Scientifica.

Daily, G. (1997). *Nature's services: Societal dependence on natural ecosystems*. Washington, DC: Island Press.

Daly, H. E., & Farley, J. (2004). *Ecological Economics: Principles And Applications*. Island Press.

De Carolis, D., Ferrari, E., & Police, A. (Eds.). (2006). *Ambiente, attività amministrativa e codificazione. Atti del primo colloquio di diritto dell'ambiente, Teramo, 29-30 April 2005*. Milan: Giuffrè.

De Groot, R. S., Wilson, M. A., & Boumans, R. M. J. (2002). A typology for the classification, description, and valuation of ecosystem functions, goods, and services. *Ecological Economics, 41*(3), 393–408.

De Leonardis, F. (2004). *L'ambiente tra i principi fondamentali della Costituzione*. Retrieved from www.federalismi.it.

De Leonardis, F. (2005). *Il principio di precauzione*. Milan: Giuffrè.

Dell'Anno, P. (2003). *Manuale di diritto ambientale*. Padua: Cedam.

Dell'Anno, P. (2004). *Principi del diritto ambientale europeo e nazionale*. Milan: Giuffrè.

Di Plinio, G., & Fimjani, P. (Eds.). (2002). *Principi di diritto ambientale*. Milan: Giuffrè.

Dietz, T., Ostrom, E., & Stern, P. C. (2003). The struggle to govern the commons. *Science, 302*, 1907–1912.

Duraiappah, A. K., et al. (2014). Managing the mismatches to provide ecosystem services for human well-being: A conceptual framework for understanding the New Commons. *Current Opinion in Environmental Sustainability, 7*, 94–100. Retrieved from www.sciencedirect.com.

European Environment Agency. (1999) *Information tools for environmental policy under conditions of complexity*. In S. Funtowicz, J. Martinez-Alier, G. Munda, & J. R. Ravetz (Eds.), Environmental issues (Series no. 9). Retrieved from http://www.eea.eu.int.

Falzea, A. (1965) *Efficacia giuridica*. In Enc. Dir. (Vol. XIV).

Falzea, A. (1996). *Introduzione alle scienze giuridiche, part I, Il concetto del diritto* (V ed.). Milano: Giuffrè.

Farì, A. (2013). *Beni e funzioni ambientali*. Naples: Jovene.

Ferrara, R. (Ed.). (2000). *La valutazione di impatto ambientale*. Padua: Cedam.

Ferrara, R. (Ed.). (2006). *La tutela dell'ambiente, in Trattato di diritto privato dell'unione europea* (Vol. XIII). Turin: Giappichelli.

Folke, C., Pritchard, L., Berkes, F., Colding, J., & Svedin, U. (2007). The problem of fit between ecosystems and institutions. *Ecology and Society, 12*(1), 30. Retrieved from http://www.ecologyandsociety.org/vol12/iss1/art30/.

Fonderico, F. (2003). La tutela dell'ambiente. In S. Cassese (Ed.), *Trattato di diritto amministrativo, t. II*. Milan: Giuffrè.

Fonderico, F. (2006). *La codificazione del diritto dell'ambiente in Italia: modelli e questioni*. In Riv. trim. dir. pubbl., no. 3.

Fonderico, F. (2008). *Ambiente (tutela dell') – Diritto Amministrativo*. In Enc. Giuridica Treccani, Agg. XVI. Rome: Istituto dell'Enciclopedia Italiana.

Fracchia, F. (2002). *Sulla configurazione giuridica unitaria dell'ambiente: art. 2 Cost. e doveri di solidarietà ambientale*. In Dir. economia, no. 2.

Fracchia, F. (2010). *Lo sviluppo sostenibile. La voce flebile dell'altro tra protezione dell'ambiente e protezione della specie umana*. Turin: Giappichelli.

Fracchia, F. (2013). *Introduzione allo studio del diritto dell'ambiente. Principi, concetti e istituti*. Naples: Editoriale Scientifica.

Georgescu Roegen, N. (1973). The entropy law and the economic problem. In H. Daly (Ed.), *Toward a steady-state economy* (pp. 37–49). San Francisco: Freeman.

Giampietro, F. (1980). *Diritto alla salubrità dell'ambiente*. Milan: Giuffré.

Giannini, M. S. (1973). *Ambiente: saggio sui diversi suoi aspetti giuridici*. In Riv. trim. dir. pubbl., no. 1.

Gomarasca, P. (2010). L'ordinamento giuridico come sistema nella prospettiva di Hans Kelsen. In L. Urbani Ulivi (Ed.), *Il pensiero sistemico come specchio di una realtà complessa* (Vol. I, p. 271 ff). Bologna: Il Mulino.

Grassi, S. (2003). *Prospettive costituzionali della tutela dell'ambiente*. In Rass. parlam., II.

Grassi, S., Cecchetti, M., & Andronio, A. (Eds.). (1999). *Ambiente e diritto*. Florence: Leo S. Olschki.

Gunderson, L. H., & Holling, C. S. (Eds.). (2002). *Panarchy: Understanding transformations in systems of humans and nature*. Washington, DC: Island Press.

Gunderson, L. H., & Pritchard, L. (Eds.). (2002). *Resilience and the behaviour of large-scale systems. Scientific Committee on Problems of the Environment (SCOPE 60)*. Washington, DC: Island Press.

Gustafsson, B. (1998). Scope and limits of the market mechanism in environmental management. *Ecological Economics, 24*, 259–274.

Habermas, J. (1996). *Between facts and norms*. Cambridge: Polity.

Habermas, J. (1998). La sfida dell'etica ecologica alla concezione antropocentrica. In S. Dellavalle (Ed.), *Per un agire ecologico. Percorso di lettura attraverso le proposte dell'etica ambientalista*. Milan: Baldini and Castoldi.

Hardin, G. (1968). The tragedy of the commons. *Science, 162*(3859), 1243–1248.

Herrero De La Fuente A. A. (2001). La Convenzione europea sul paesaggio, Riv. Giur. amb., 893 ss.

Hester, R. E., & Harrison, R. M. (Eds.). (2010). *Ecosystem services*. Cambridge: Royal Society of Chemistry.

Holland, J. H. (1992). *Adaptation in natural and artificial systems: An introductory analysis with applications to biology, control, and artificial intelligence*. Cambridge: The MIT Press.

Holland, J. H. (1996). *Hidden order: How adaptation builds complexity*. Cambridge: Perseus Books.

Holland, J. H. (2000). *Emergence: From chaos to order*. Oxford: Oxford University Press.

Holland, J. H. (2013). *Signals and boundaries. Building blocks for complex adaptive systems*. Cambridge: The MIT Press.

Holland, J. H. (2014). *Complexity: A very short introduction*. Oxford: Oxford University Press.

Holling, C. S. (1973). Resilience and stability of ecological systems. *Annual Review of Ecological Systems, 4*, 1–23.

Holling, C. S. (1988). Paradigmi della funzione e della struttura ecologiche. In M. Ceruti & E. Laszlo (Eds.), *Physis: abitare la terra*. Milan: Feltrinelli.

Iovino, S. (2004). *Filosofie dell'ambiente. Natura, etica, società*. Rome: Carocci.

Leakey, R., & Lewin, R. (1995). *The sixth extinction: Patterns of life and the future of humankind*. New York: Doubleday.

Lee, K. N. (1993a). Greed, scale mismatch, and learning. *Ecological Applications, 3*(4), 560–564.

Lee, K. N. (1993b). *Compass and gyroscope. Integrating science and politics for the environment*. Washington, DC: Island Press.

Levin, S. (1999). *Fragile Dominion. Complexity and the commons*. Cambridge: Helix Books.

Licata, I. (2013). Incertezza. Un approccio sistemico. In L. Urbani Ulivi (Ed.), *Strutture di mondo. Il pensiero sistemico come specchio di una realtà complessa* (Vol. II). Bologna: Il Mulino.

Losano, M. G. (2002). *Sistema e struttura nel diritto* (3 vols.) Milan: Giuffrè.

Maddalena, P. (1987). *Il danno all'ambiente tra giudice civile e giudice contabile*. In Riv. crit. dir. priv., no. 3.

Maddalena, P. (1990). *Il diritto all'ambiente ed i diritti dell'ambiente nella costruzione della teoria del risarcimento del danno pubblico ambientale*. In Riv. giur. ambiente, no. 3.

Maddalena, P. (2010). *L'interpretazione dell'art. 117 e dell'art. 118 della Costituzione secondo la recente giurisprudenza costituzionale in tema di tutela e di fruizione dell'ambiente*. In Federalismi.it, no. 10.

Maddalena, P. (2013). *Per una teoria dei beni comuni*. In "Micromega", no. 9.

Marella, R. (2012). *Oltre il pubblico e il privato. Per un diritto dei beni comuni*. Verona: Ombre corte.

Marten, G. G. (2001). *Human ecology, basic concepts for sustainable development*. London: Earthscan Publications.

Mezzetti, L. (2002). *Manuale di diritto ambientale*. Padua: Cedam.

Millennium Ecosystem Assessment. (2003). *Ecosystems and human well-being: A framework for assessment*. Washington, DC: Island Press Retrieved from www.millenniumassessment.org.

Montini, M. (2017). *Riflessioni critiche sull'accordo di Parigi sui cambiamenti climatici*. In Rivista di Diritto Internazionale, fasc. 3.

Morbidelli, G. (1996). *Il regime amministrativo speciale dell'ambiente*. In AA.VV. *Scritti in onore di Alberto Predieri*, t. II, Milan, Giuffrè.

Nespor, S. (2016). *La lunga marcia per un accordo globale sul clima: dal Protocollo di Kyoto all'Accordo di Parigi*. In Rivista trimestrale di diritto pubblico.

Ostrom, E. (1990). *Governing the commons. The evolution of institutions for collective action*. Cambridge: Cambridge University Press.

Ostrom, E. (1996). *Né mercato né stato nella gestione delle risorse collettive*. In "La Questione Agraria", no. 64.

Ostrom, E. (1998). Scales, polycentricity, and incentives: Designing complexity to govern complexity. In D. L. Guruswamy & J. A. McNeely (Eds.), *Protection of global biodiversity: Converging strategies*. Durham: Duke University Press.

Ostrom, E. (2005). *Understanding institutional diversity*. Princeton: Princeton University Press.
Pearce, D. W., & Turner, R. K. (1990). *Economics of natural resources and the environment*. Hemel Hempstead: Harvester Wheatsheaf.
Pericu, G. 1987 *Ambiente (tutela dell') nel diritto amministrativo*. In Digesto (Disc. Pubbl., vol. I), Turin.
Peters, G. P. (2008). From production-based to consumption-based national emission inventories. *Ecological Economics, 65*, 13–23.
Porena, D. (2009). *La protezione dell'ambiente tra Costituzione italiana e "Costituzione globale"*. Turin: Giappichelli.
Postiglione, A. (1985). *Ambiente: suo significato giuridico unitario*. In Riv. trim. dir. pubbl., no.1.
Pototschnig, U. (1970). *Strumenti giuridici per la difesa della natura*. In Foro Amm., no. 46 (p. III).
Predieri, A. (1969). *Significato della norma costituzionale sulla tutela del paesaggio, in Studi per il ventesimo anniversario dell'Assemblea Costituente* (Vol. II). Vallecchi: Florence.
Predieri, A. (1981). *Paesaggio*. In Enc. dir., Milano (Vol. XXXI).
Prigogine, I. (1988). L'origine della complessità. In M. Ceruti & E. Laszlo (Eds.), *Physis: abitare la terra*. Milan: Feltrinelli.
Prigogine, I., & Stengers, I. (1989). *Tra il tempo e l'eternità*. Turin: Bollati Boringhieri.
Prigogine, I., & Stengers, I. (1999). *La nuova alleanza* (3rd ed.). Turin: Einaudi.
Rajamani, L. (2016). The 2015 Paris agreement: Interplay between hard, soft and non-obligations. *Journal of Environmental Law, 28/2*, 337.
Rammel, C., & Van den Bergh, J. (2003). Evolutionary policies for sustainable development: Adaptive flexibility and risk minimising. *Ecological Economics, 47*(2.3), 121–133.
Ramos-Martin, J. (2003). Empiricism in ecological economics: A perspective from complex systems theory. *Ecological Economics, 46*(3), 387–398.
Rapisarda Sassoon, C. (Ed.). (2002). *Manuale delle leggi ambientali*. Milan: Giuffrè.
Renna, M. (2012). *I principi in materia di tutela dell'ambiente*. In Riv. Quadr. di Dir. Ambiente, 1-2.
Romano, S. (1946). *L'ordinamento giuridico* (II ed.). Florence: Sansoni.
Ross Ashby, W. (1956). *An introduction to cybernetics*. London: Chapman & Hall. Retrieved from http://pcp.vub.ac.be/books/IntroCyb.pdf.
Ruhl, J. B., Kraft, S. E., & Lant, C. L. (2007). *The law and policy of ecosystem services*. Washington: Island Press.
Scoca, F. G. (1979). *Contributo sul tema della fattispecie precettiva*. Perugia: Libreria Editrice Universitaria.
Scoca, F.G. (1993). *Osservazioni sugli strumenti giuridici di tutela dell'ambiente*. In Diritto e società, no. 3.
Simon, H. A. (1962). The architecture of complexity. *Proceedings of the American Philosophical Society, 106*(6), 467–482.
Simon, H. (1988). *L'architettura della complessità, in Le scienze dell'artificiale*. Il Mulino: Bologna.
Sorace, D. (1999) *Considerazioni conclusive*. In Grassi, Cecchetti and Andronio.
Stiglitz, E. (2000). *Economics of the public sector*. New York: W. W. Norton.
Tallacchini, M. C. (1996). *Diritto per la natura. Ecologia e filosofia del diritto*. Turin: Giappichelli.
Vv.Aa., The Paris Agreement (2016) Review of European, Comparative and International Environmental Law.
Tiezzi, E. (1988). Tempi storici e tempi biologici. In M. Ceruti & E. Laszlo (Eds.), *Physis: abitare la terra*. Milan: Feltrinelli.
Troper, M. (2001). *La théorie du droit, le droit, l'état*. Paris: PUF.
Weber, C. L., Peters, G. P., Guan, D., & Hubacek, K. (2008). The contribution of Chinese exports to climate change. *Energy Policy, 36*(9), 3572–3577.
Weisbrod, B. (1964). Collective consumption services of individual consumption goods. *Quarterly Journal of Economics, 78*, 471–477.

Weizsacker, E., & Weizsacker, C. (1988). Come vivere con gli errori? Il valore evolutivo degli errori. In M. Ceruti & E. Laszlo (Eds.), *Physis: abitare la terra*. Milan: Feltrinelli.

Wilson, J. A. (2006). Matching social and ecological systems in complex ocean fisheries. *Ecology and Society, 11*, 9. Retrieved from www.ecologyandsociety.org/.

Young, O. R. (2002). *The institutional dimensions of environmental change. Fit, interplay, and scale*. Cambridge: The MIT Press.

Zito, A. (2006). I limiti dell'antropocentrismo ambientale e la necessità del suo superamento nella prospettiva della tutela dell'ecosistema. In D. De Carolis, E. Ferrari, & A. Police (Eds.), *Ambiente, attività amministrativa e codificazione. Atti del primo colloquio di diritto dell'ambiente, Teramo, 29-30 April* (Vol. 2005). Milan: Giuffrè.

The Living Body as a Model of Systemic Organization in Ancient Thinking

Elisabetta Matelli

Abstract Analyzing Homer and Aristotle, the Author faces the ancient Greek origin of the organicist model (introduced since 1920 in system theory) presenting its features. In Homer there is still no term to indicate the living body as a whole, but is present the idea of a principle capable of giving "shape" (*eidos*) to body elements and to counteract the natural tendency to disintegration: the soul (*psyché*). Only with Aristotle the living body begins to be understood as "organism," thanks to a *hylomorphic* and non-dualistic vision of the relationship of the soul with matter, which explains the living organism. The soul itself, in Aristotle, has the characteristics of a system. From this analysis, the organicist model seems to be enriched by the indispensable notion of "form" which, in turn, calls for the need for an efficient cause outside the system.

1 Introduction

The general theory of systems expressly recognizes its affinity with the organicist concept in biology: "Any organism is a system, that is, a dynamic order of parts and processes standing in mutual interaction."[1] Many contemporary organizational systems—and here I shall limit myself to mentioning a few emergent points, such as social, political, economic environment, and within this latter, the corporate environment—have, for about a century, recognized a very valuable reference in systems theory, even though based on different models, not last of which is that of mechanization, proposed as an alternative to the biological model.[2] The value of the organicist model of organization emerges in particular

[1] Bertalanffy (1969: 208).

[2] Morgan (1986): on organizations as "machines" [19–38]; on organizations as organisms [39–76];

E. Matelli (✉)
Dipartimento di Filologia classica, Papirologia e Linguistica storica, Università Cattolica del Sacro Cuore, Milan, Italy
e-mail: elisabetta.matelli@unicatt.it

L. Urbani Ulivi (ed.), *The Systemic Turn in Human and Natural Sciences*, Contemporary Systems Thinking, https://doi.org/10.1007/978-3-030-00725-6_8

for its capacity, on the theoretical level as well as the experimental, to provide well-functioning paradigms in situations of variability and indeterminacy, such as the majority of human experiences and, not last, also the explanation of matter and the universe by quantum physics.[3]

As a scholar of classical philology, I begin from Homer in investigating how the first awareness of the unitary functioning of the biological body constituted of parts was formed, and how his unity could then later become a paradigm for other complex realities.

Researches into classical antiquity have developed the theme of corporeality from different points of view (lexical, mythological, religious, medical, philosophical, literary, figurative), and it is not possible to take into account the details here, within a framework of extreme synthesis in regard to the Homeric origins and Aristotle's fundamental theory, elaborated in the fourth century BC.[4] Significant contributions regarding the idea of the body as an organic unity in the classical texts have been developed especially by the researches on Christian theological texts dealing with this theme. They recognize that in early Christian thinking, and particularly in the reflections of St. Paul, the centrality of the *body*, as well as the articulation of its theological values (first and foremost, those of the "body" of Christ and consequently those of the "body" of the Church), makes use of Greek and Roman lexical and semantic categories relating to the Judaic culture,[5] bringing forth from this ground an absolutely new and original[6] theological and anthropological model, of which I will make mention in the conclusion.

---

on organizations on the model of cerebral functioning [77–109]; the potentialities and limits of mechanistic, organicist, and cerebral models are considered at pp. 33–38, 71–76, and 105–109, respectively. With a historical-descriptive approach, Scott (1992) studies and contrasts the organization of a system according to the "rational" model [29–50] with that of one according to the "natural" model, proper to "dynamic" systems [51–75]. Both of these systems can be "open" or "closed" [76–94] and can combine with each other [95–124]. The interesting conclusion of this study is that it is not possible to define *a priori* the most effective type of organization [342–362]. The complexity of the various components of an organizational system, along with the variability and the indeterminacy of situations, even when provided with highly detailed information, demonstrates our inability to control the absolute effectiveness of the functioning of individual systems given that, in contingent situations, some degree of unpredictability always remains. Even in the best of cases, therefore, something always seems to escape our rational domain. For these reasons, systems always have histories that are more or less limited in time, or they are subject to transformations that change their identity.

[3] Del Giudice (2010: 47–70), Villani (2010: 71–89), Bertolaso (2013, 143–169).

[4] I would like to point out some of the studies that have set out such issues, first of all, the body–soul relationship, Rohde (1921); Hirzel (1914); Böhme (1929), Snell (1955: 15–37); Meyer (2008); Jaeger (1953: 88–106 and notes at p. 261–264); Onians (1998); Jahn (1987); Schmitt (1990); Zavalloni (1990).

[5] For just a few bibliographical references: Adinolfi (1963: 333–342); Bartolomei (1984); Bellia and Garribba (2011), especially Bellia (2011), Jossa (2011), and Pitta (2011: 75–77).

[6] Cf. the contributions of Schütz and Wibbing (2010: 1279–1280) and Wibbing (2010: 1281–1285); Freedman (1992: 767–772: 768); Pitta (2011: 76–77); Schweizer (1981, cols. 609–790), Viagulamuthu (2002), Sichkaryk (2011).

2 Homer

Any inquiry into the key concepts of Western culture can only start from Homer.[7] In *The Iliad* and *The Odyssey*, there is no term relating to the idea of "body" that coincides with the modern meaning of the word in the sense of "organism"[8]: the neutral term *sôma* will be used here in reference to bodies of animals or men, always and only[9] in the sense of "carcass"/"cadaver."[10]

Despite the absence of a specific term for making reference to the living body, the wealth of the lexicon relating to the parts of the human organism is surprising. The battles or Odysseus's peril-filled return to his homeland, in fact, cause the human physicality of the protagonists to emerge, mortal heroes who face trials and armed struggles, enduring or causing wounds that were, for the most part, fatal: all these highlight not only an awareness of the physiology and anatomy of the human body, but also the connection between emotive tensions and corporeal organs. The poetic majesty of the Homeric compositions lies primarily in that universality which emerges from the extreme precision of the nomenclature and anatomic descriptions. The intention is poetic, and one certainly ought not seek scientific information therein, especially considering that these poems are recognized as having a compositional stratification that took place over a long period of time; however, the culture underlying even the most archaic formulary lexicon appears to possess notions of human anatomy that are not superficial.[11]

3 Physical Anatomy

In Book IV of *The Iliad*, the poet tells about the tip of the bronze lance of Nestor's son Antilochus penetrating *the bone* (*to osteon*) *of the forehead* (*to metôpon*) of Echepolus, son of Thalysius, describing his death as a darkness that veiled his *eyes* (*tô osse*)[12]; in Book XIV the anatomical lexicon is rather detailed in recounting how the Boeotian Peneleos killed the Trojan Ilioneus, son of Phorbas: "he wounded him in the depths *of the eye* (*ho ophtalmos*), under *the eyebrows* (*hê ophrys*), tearing out

[7] For *The Iliad* and *The Odyssey*, unless otherwise indicated, my translation is literal, rather than literary.

[8] Lehrs (1964: 86–87 and 160) contains the observations by the grammarian Aristarchus on the notion of *sôma* in the Homeric poems, according to whom "*sôma* apud Homerum dicitur tantum de *cadavere*." Cf. Snell (1955: 19–22), Jaeger (1953: 89).

[9] The interpretation of *Il.* 3.23 can only be dubious.

[10] In these verses: *Il.* 3.23, 7.79, 18.161, 22.342, 23.169; *Od.* 11.53, 12.67, 24.187. The lexicon of corporeality is studied specifically by Vivante (1955: 43–44).

[11] Urso (1997).

[12] *Il.* 4. 460–461. Cf. Urso's observations (1997: 26), which acknowledge the reference to the frontal bone, but also that "it is not possible from the text to infer whether or not it was known that the cranium is made up of various articulated bones rather than being one single bone."

the eyeball (*hê glênê*)[13]; the spear went right *through the eye* to *the nape of the neck* (*to inion*); and he fell, stretching out *both hands* (*tô cheire*); and Peneleos, after pulling out the sharpened sword, smote him full downward on the *neck* (*ho auchên*), sweeping *the head* (*to karê*) with its helmet away to the ground, the robust staff still stuck *in his eye*."[14] The description of the death of the Trojan Erymas in Book XVI shows how the connection was obvious among the various organs like the mouth and teeth, the upper part of the pharynx, the ocular cavities, and the bony base of the encephalon: "Idomeneus wounded Erymas in the *mouth* (*to stoma*) with his pitiless bronze. The spearhead passed on into and through *the mouth* beneath *the brain* [*ho enkephalos (myelos)*], smashing the white bones (*ta ostea*). His teeth (*hoi odontes*) were dashed out, and blood (*to haima*) filled both *his eyes*, gurgling forth from *his mouth* and *nostrils* (*hai rhînes*) as his *mouth* gaped open. Death's black cloud enveloped him."[15] In Book V, there is a description of the anatomy of the oral cavity within the cranial structure, when Achaean Meges, son of Phyleus, strikes the Trojan Pedaeus, the bastard son of Antenor: "Meges Phyleides, the master spearman, closing with him, struck him in *the head* (*hê kephalê*), *at the nape* with the pointed spear. Straight through to his teeth, the bronze cut off his *tongue's root* (*hê glôssa*). Biting the cold bronze with *his teeth* he fell into the dust."[16] And we observe no less precision, although with different details,[17] in the narration of Diomedes's blow that killed Pandarus: "Athena guided the weapon to cleave his nose (*hê rhis*) beside *the eye* and shatter his white *teeth*; the inflexible bronze spearhead severed his *tongue* at the base, then plowing on came out beneath the tip of his *chin* (*ho anthereôn*),"[18] or of that of Hector, who struck the Achaean Coeranus under *the jaw* (*ho gnathmos*) and *the ear* (*to ouas*), cutting off his *tongue*.[19]

In Book XX, when Achilles cuts off Deucalion's head, the connection between the cranial region and that of the vertebra containing the marrow appears explicit: "Achilles with his dagger cut his *neck* and knocked both *head* and helmet far away; *the marrow* (*ho myelos*, perhaps corresponding, here, to the *medulla oblongata* or bulb) squirted forth from the vertebrae (*ta sphondylia*)."[20] Shortly before, Achilles had struck Tros, son of Alastor, in *the liver* (*to hêpar*), with his dagger, and as the liver squirted, it spilled forth *bile*, called "black blood"[21] (*to melan haima*, but

[13] Observe Urso (1997: 26): "The term *glênê* (= pupil) will be used posteriorly with the meaning of bone hollow, distinct from *kotylê*, which indicates a deeper bone cavity. This suggests perhaps that in the Homeric age it was known that the pupil was a cavity (or hole) of the eye, but it certainly indicates that the term is used precisely to indicate that structure and not applied generically."

[14] *Il.* 14.493–499.

[15] *Il.* 16.345–350.

[16] *Il.* 5.72–75.

[17] Urso (1997: 27) notes that the blows delivered to the laterocervical region, in severing the external carotid artery, were nearly always fatal for heroes of *The Iliad*.

[18] *Il.* 5. 290–293.

[19] *Il.* 17. 617–618, cf. 13.671–672 and 16.606–607.

[20] *Il.* 20.481–483.

[21] *Il.* 20.469–471.

elsewhere called *cholon*[22]) and with a bronze spear pierced through Deucalion's *arm* (*hê cheir*), at the place where the *tendons* (*hoi tenontes*) of the elbow (*ho ankôn*) meet.[23] The temporal bone is recognized by the name *krotaphos*, and it is passed through by Odysseus's spear which pierces Democoon, the natural son of Priam[24] and by that of Achilles, who strikes Demoleon, son of Antenor[25]; otherwise, the plural *hoi krotaphoi* is used to refer to the temples, which can turn white.[26] Diomedes struck Astynoos in *the breast* (*ho mazos*) and Hypeiron "*at the shoulder on the collarbone* (*ho ômos*)," so that "the shoulder was separated from the nape of the neck and from the back (*to nôton*)."[27] Meriones, in pursuing Phereklos, caught up with him and "hit him on the right *buttock* (*ho gloutos*), and straight forward through the *bladder* (*hê kystis*) the tip arrived to the bone."[28] The attack the Diomedes unleashes against Aeneas, also, is described with extreme anatomic precision: "with it he caught Aeneas on the *hip* (*to ischion*), where the *thigh* (*ho mêros*) curves to form the *hip*: they call it *cotila* (*hê kotylê*). He crushed the *cotila*, and broke *both the sinews*, while the jagged stone tore away the *flesh* (*to rhînon*)."[29]

In numerous other scenes, the spears attack the enemy, striking them in the heart (*hê kradiê*[30]/*to êtor*[31]), the belly (*hê gastêr*[32] or *hê nêdys*[33]), the bowels (*hai cholades*),[34] the intestines (*ta entera*),[35] the navel (*omphalos*),[36] the genital organs (*ta aidoia*),[37] the sternum (*sternon*) and the chest (*stêthos*),[38] the last vertebra (called *ho astragalos*),[39] the aorta (*hê phleps*),[40] the trachea (*ho aspharagos*),[41] the lung

[22] *Il.* 16.203.

[23] *Il.* 20.478–480.

[24] *Il.* 4. 501–502.

[25] *Il.* 20.397–400.

[26] *Il.* 8.518, 13.188 and 805, 15.609 and 648, 16.104, 18.611; *Od.* 11.319, 18.378, 22.102.

[27] *Il.* 5.146–147.

[28] *Il.* 5.65–67.

[29] *Il.* 5.305–308.

[30] *Il.* 13.442.

[31] *Il.* 1.188–189 and 3.31.

[32] *Il.* 4.531, 5.539, 5.616, 13.372, 13.398 and 506, 16.163 and 465, 17.313 and 519, 21.180.

[33] *Il.* 13.290.

[34] *Il.* 4.526, 21.181.

[35] *Il.* 13.507, 14.517, 17.314–15, 20. 418 and 420.

[36] *Il.* 4.525, 13.568, 20.416, 21.180.

[37] *Il.* 13.568.

[38] The *sternum*: *Il.* 4.528 and 530, 13.290, 15.542, 16.312 and 400. The *chest*: *Il.* 4.108 and 480, 5.19, 41, 57, 317 and 346, 8.121, 303, 313 and 326, 11.108 and 144, 13.186 and 586, 15.420, 523, 577 and 650, 16.597, 17.606. Cf. *Od.* 9.301, 22.82.

[39] *Il.* 14.466. Cf. *Od.* 10.560 and 11.65.

[40] *Il.* 13.546.

[41] *Il.* 22.327–329 (this is the episode in which Achilles strikes a deadly blow to Hector, but never pierces the trachea, so that the Trojan hero is able to utter some final words). Urso (1997: 28) observes that "the clarification in the text that Hector could still articulate words after having been

(*pneumon*),[42] the diaphragm (*hai prapides*[43] or *hai phrenes*[44]), and still other limbs. However, this is not the place to expose in detail and with the necessary critical method the anatomical references in *The Iliad* and *The Odyssey*. By mentioning some examples, I intended mainly to recall the precise and conscious way Homeric poetry describes the various parts making up the human body.

4 Connections Among Parts of the Body and Emotions

Injuries and deaths, especially on the battlefields, made it possible to observe not only the connections between the various limbs, but also to recognize the connection between the physical organs and the emotions and feelings of the heroes. Some limbs are recognized as the seats in which psychic faculties originate, and from which the vital principle comes to an end at time of death.[45] Just a few examples:

Hai phrenes,[46] the diaphragm, which "trembles within,"[47] "envelops the heart,"[48] or "encloses the liver,"[49] is this seat both of eros[50] and of the other passions (primarily anger[51]), and of the state of consciousness and rational thoughts.[52] Therefore, by way of metonymy, the term *phrenes* can also indicate the "soul," the "mind," or even "thoughts" themselves.[53] The diaphragm is also indicated by another feminine plural term, *hai prapides*, which refers to the physical organ: but in the moving encounter between Achilles and Priam, it indicates, together with the knees (*ta gyia*), the seat of the emotional state (*himeros*).[54] In a recurring formula, it expresses ability, good sense.[55]

---

struck raises the question of whether the term employed, *aspharagos* (= trachea), refers to the larynx or to the trachea-larynx complex. The fact that the term *larynx* (= larynx) is never used in Homer would support this hypothesis."

42 *Il.* 4.528.

43 *Il.* 11.579, 13.412. 17.349.

44 *Il.* 16.481, cf. 504.

45 Onians (1998: 44–62); Jahn (1987: 9–27).

46 Onians (1998: 93–122); Jahn (1987: 17–19).

47 *Il.* 10.10.

48 *Il.* 16.481.

49 *Il.* 11.579, 13.412, 17.349; *Od.* 9.301.

50 *Il.* 3.442, 14.294.

51 *Il.* 1.103, 4.661–662.

52 *Il.* 24.40: "Achilles, who does not have sound reason" and *Od.* 18.215. In *Il.* 23.103–104, the formula is in reference to the absence of "mind" in the *psychai* that are in Hades, as in *The Iliad* 24.201: "Alas, where did your mind go?" Cf. *Od.* 10.493, concerning the rational mind, which only the soothsayer Tiresias conserved, by gift of Persephone, even in Hades.

53 *Od.* 11.367.

54 *Il.* 24.51.

55 *Il.* 1.608, 18.380 and 482, 20.12; *Od.* 7.92.

Hê kradiê, the heart, "swells with anger,"[56] "sees death,"[57] and "inwardly, the thought of death palpitates very strongly, and the teeth chatter (*ta odonta*)"[58]; furthermore, the heart "is uncertain,"[59] "is astonished,"[60] "is agitated,"[61] "consumes itself,"[62] "thrusts" and "exhorts,"[63] sometimes it obeys,[64] other times "it commands"[65] and even "howls."[66] The heart can be "harder than a stone,"[67] or "of iron."[68] *Chalcheon êtor* expresses, on the other hand, a "heart of bronze," impassive.[69] Odysseus speaks to his own heart, remonstrating that it knows well how to endure terrible things.[70] When Andromeda foretells the misfortunes of Priam's children, her heart (*êtor*) "beats (*palletai*) in her chest (*stêthesi*) up to her throat (*ana stoma*), with the knees (*gouna*) below, rigid."[71]

The heart (*êtor*) is the seat of reason in *The Iliad* 1.188. A recurring formula in *The Iliad* and *The Odyssey* describes how strong emotions cause *the heart and knees* (*autou lyto gounata kai philon êtor*) to fail, with reference to both the emotive/sentimental[72] sphere and the erotic.[73]

5 The Part Expresses the Whole

The knees are indicated not only by *ta gounata*, but also by *ta gyia*; this latter term, always given in the plural, has more complex semantic values: Poseidon and Athena reinvigorate their heroes by making agile "the knees (*gyia*), feet, and arms,"[74] but the term is often utilized in formulas that express the effect of fear or terror, as in our

[56] *Il.* 9.646; *Od.* 18.348.

[57] *Od.* 5.389: the verb is *protiossomai*.

[58] *Il.* 13.282–83.

[59] *Il.* 1.188–189: the heart is expressed by *to êtor* and the verb is *mermêrizô*.

[60] *Il.* 3.31: the heart is expressed by *to êtor* and the verb is *kataplêssô*.

[61] *Od.* 4.427, 10.309: the verb is *porphyrô*.

[62] *Od.* 4.467.

[63] *Od.* 14.517, 15.339, 16.81, 21.198 e 342: the verb is *keleuô*. In *Il.* 10.220 and 319, *Od.* 18.61 the heart (*hê kradiê*) is in hendiadys with ire (*ho thymos*) and the verb is *otrynô*.

[64] *Od.* 20.22–24: literally "remains in obedience," as contrasted with physical agitation.

[65] *Od.* 15.395: the verb is *anôgô*.

[66] *Od.* 20.13 and 16: the verb is *hylaktô*.

[67] *Od.* 23.103 (*kradiê stereôterê lithoio*).

[68] *Od.* 4.293 (*kradiê sidêreê*).

[69] *Il.* 2.490.

[70] *Od.* 20.17–18.

[71] *Il.* 22.452–453.

[72] *Il.* 21.114, 21.425; *Od.* 4.703, 5.297 and 406, 22.68 and 147, 24.345, 24.345.

[73] *Od.* 18.211 and 24.381.

[74] *Il.* 13.61, 23.772.

expression "the knees tremble,"[75] or it indicates the sudden arrival of tiredness, exhaustion that makes the limbs weak (owing to an effort, but also owing to lack of care or old age).[76] *Gyia lyein* "melt/let go/bend the knees" expresses the act of dying, while the *thymos* (the vital spirit, which breathes) abandons the limbs.[77] Formulas of the type "Hector struck Eioneus with his pointed spear, at the neck, beneath the helmet's well-worked bronze, and his knees buckled (*lynto de gyia*),"[78] or "he fell to earth from the chariot, and his knees buckled (*lynto de gyia*),"[79] to express—through the bending of the knees—the death of heroes, make it possible to understand how the knees, in these cases, represent *the part* for the *whole*, or (with the poetic artifice we call *synecdoche*) the failing of the *entire body*, even if, for this latter concept, there is not a specific term. Onians demonstrates, not by chance, that the knees and the femurs, analogously with the head, were considered as seats of the vital spirit.[80]

In the Homeric poems, other terms as well seem to be utilized to indicate the whole of the body by means of one part. For example, in *The Iliad* 5.348, "no one will be able to keep the dogs away from your body," the idea of the body is expressed by *hê kephalê*, literally "from the head," a part of the body in which the presence of the entire person is perceived.[81] In *The Odyssey* 11.29, the supplication for the body of the dead refers to their "heads" (*ta karêna*), and in *The Iliad*—with formulary language—the "tremendous sorrow" of the heroes "penetrates the soul (*ho thymos*) and the heart (*hê kradiê*)."[82]

The consciousness of the unitary connection between the different organs and the feelings of the hero emerges: limbs that are different and distant from one another respond together to fear, each in its own specific way, in verses such as "my heart (*êtor*) is not in peace, but I am upset, my heart (*kradiê*) leaps out of my chest (*exo stêtheôn*) and, below, my well-made knees (*phaidima gyia*) are trembling."[83]

Is the unitary notion of the *living body* truly lacking in Homer, then, as we would be led to believe by the absence of a specific term?[84] It does not seem possible for the answer to be affirmative. On the contrary, the multiplicity of physical components will receive unity from the presence of a unitary and individual *form* of theirs, which goes into Hades after the death of the physical body, called both *eidos* and

---

[75] *Il.* 3.34, 7.215, 8.452, 10.95 and 390, 14.506, 20.44, 22.448, 24.170; *Od.* 11.527, 18.88 and 341.

[76] *Il.* 4.230, 5.811, 7.6, 13.85, 19.165 and 169, 23.63, 23.627 (for old age); *Od.* 1.192, 8.233, 12.279.

[77] *Il.* 4.469, 6.27, 7.12 and 16, 11.240 and 260, 15.435 and 581, 16.312, 341, 400, 465, 805, 17.524, 18.31, 21.406, 23.691; *Od.* 18.238 and 242.

[78] *Il.* 7.11–12.

[79] *Il.* 7.16.

[80] Onians (1998: 174–186).

[81] Onians (1998: 98–100).

[82] *Il.* 2.171, 8.147, 10.220, 10.244, 10.319, 15.208, 16.52.

[83] *Il.* 10.93–95. Cf. Schmitt's *status quaestionis* (1990: 115–116, 271–272 note 352).

[84] For these observations, see Hirzel (1914: 6–7), Snell (1955: 15–37), Schweizer (1981, col. 612 note 3), Freedman (1992: 768).

psychê, and of a particular state of individual consciousness, which synthesizes in itself different psychic and cognitive phenomena, called *thymos*,[85] common also to animals,[86] which is made volatile at the moment of disintegration of the physical elements coinciding with death.

We observe that, in addition to *gyia*, two other neutral, plural terms, *melea* e *rhethea*[87] (literally, in their own right, "limbs") are often used in formulas that—even while designating parts, namely the "limbs"—require that these be interpreted in reference to the entire human body, always using the logical (and poetic) artifice of synecdoche.[88] A typical formula is "immediately the spirit (*ho thymos*) left the limbs (*apo meleôn*), and the horrible darkness took him."[89] Analogously, we have a reference to the entire body when the poet describes the youthful vigor "of the limbs,"[90] or the sweat that flows abundantly from "the limbs,"[91] or that, owing to fear, "the flesh shivered along *the limbs* [in fact, Calzecchi Onesti translated these words into Italian "per tutto il corpo" ("all along the body")]."[92]

Ho chrôs, literally "skin," indicates "the body as it is perceived or felt in terms of color or to the touch."[93] *Demas*, "physical structure"[94]; *phyê*, "shape"[95]; and *eidos*, "form" / "aspect"[96] are the terms that express the idea of "corporeality" in numerous verses.[97]

The neutral term *demas*, even if translated, in many cases, as "physical aspect," literally indicates "bodily structure," or also "stature." It is interesting to observe that this lemma derives from the verb *demô* ("I build") and that the same root, with the vocalism *o*, recurs in *domos* "domicile"/"house." This term, as I see it, prepares

[85] Cf. below, note 104.

[86] Animals as well possess *thymos* and give it off at the moment of death: *Il.* 16.469 (death of a horse), and 23.880 (death of a dove); *Od.* 10.163 (death of a deer), 19.454 (death of a wild boar).

[87] Vivante (1955: 30 note 1 and 40–42).

[88] *Il.* 7.131, 13.671–72, 16.607, 23.191 (reference to the limbs of Hector's corpse, protected by Apollo with a cloud, so that the sun's heat would not desiccate the skin covering them), 23.880 (referring to the body of a slain dove); *Od.* 11.201, 15.354.

[89] *Il.* 13.671–72, 16.607.

[90] *Il.* 11.669 and *Od.* 11.394, 13.398, 21.283.

[91] *Il.* 16.110, 23.689 and *Od.* 11.600.

[92] *Od.* 18.77.

[93] Snell (1955: 19–20); Vivante (1955: 42).

[94] Again as accusative of relation, *Il.* 1.115, 5.801, 8.305, 13.45, 17.323 and 555, 21.285, 22. 227, 24.376; *Od.* 2.268, 2.401, 3.468, 4.796, 5.212–213, 7.210, 8.14, 8.116, 8.194, 10.240, 11.469, 13.222, 13.288, 14.177, 16.157, 16.174, 17.307, 17.313, 18.251, 19.124, 19.381, 20.31, 20.194, 22.206, 23.163, 24.17, 24.503, 24.548. In *Il.* 11.596 and 13.673, 17.366, 18.1 *demas pyros* "with an aspect like that of fire (*demas pyros*)" recurs as an adverbial formula.

[95] *Il.* 1.115, 2.58, 3.208, 22.370; *Od.* 5.212, 6.16 and 152, 7.210, 8.134 and 163.

[96] Again as accusative of relation, cf. *Il.* 2.58, 2.715, 3.39, 3.45, 3.55, 3.124, 3.224, 5.787, 6.252, 8.228, 10.316, 3.365, 13.378, 13.769, 17.142, 17.279, 21.316, 22.370, 24.376; *Od.* 4.14, 4.264, 5.213, 5.217, 6.16, 6.152, 7.57, 8.116, 8.169, 8.174 e176, 11.337, 11.469, 11.550, 14.177, 18.4, 18.249, 18.251, 19.124, 20.71, 24.17, 24.253, 24.374.

[97] Vivante (1955: 30 note 1 and 44–47).

the way for the successive and more complex idea of the *organization* of the limbs of the body in a manner similar to the structure of the building.[98]

Eidos, etymologically linked to the verbs "to see"/"to appear,"[99] expresses the idea of the *form* that appears to the sight and that makes it possible to identify individuality. This latter is a term important for the evolution of the notion of the *body* as matter *endowed with form*, where *form* is the principle of unification for the parts that otherwise are disunited; *eidos* is a concept destined for successive fundamental semantic evolutions for the development of the first ancient philosophical reflections (which, around this idea, constructed entire systems of interpretation of reality)[100] and a pivotal point for the first comprehension of the functioning of the bodily members as an organism.

6 Psychê

To Homer's anatomical references to the human body we must add a few considerations regarding the term *psychê* (in itself a keyword for understanding the living body). This does not yet express the successive meanings of "soul" or of "spirit."[101] Nonetheless, especially when it is used in hendiadys with *menos* ("strength/vigor"),[102] *aiôn* ("life"),[103] or *thymos* ("mental and sensory spirit"),[104] it indicates the vital element that flows out from the limbs of a dying person, without which, the body is nothing other than an empty container, destined to the disintegration of its parts.[105]

---

[98] Chantraine (2009: 250–251), *s.v. demô*; Beekes (2010: 314–315), *s.v. demô*.

[99] The root, i.e., **weid*, expresses the idea of "to see" present in the Greek verbs *idein* ("to see") and *oida* ("to know") and in the Sanskrit *vedas* "knowledge" (Chantraine (2009: 302), *s.v. eidos*; Beekes (2010: 379–380), *s.v. eidomai*).

[100] I would point out, indicatively, these synthesis studies, mainly devoted to *eidos* in the early naturalist philosophers, in Plato and in Aristotle: Motte et al. (2003); Fronterotta and Leszl (2005).

[101] The nineteenth-century work of Rohde (1921) was fundamental to this topic, which opened a further complex critical debate, resumed and summarized by Otto (1923: 5–15) and by Jaeger (1953: 89–106); Jahn (1987: 124–151) offers a reasoned *status quaestionis* of the interpretations of the soul/spirit in the nineteenth- and twentieth-century studies with a bibliography to which the reader is referred. Also Böhme (1929); Snell (1955: 15–37); Onians (1998: 93–122); Schmitt (1990).

[102] *Il.* 5.296 = 8.123 and 315.

[103] *Il.* 16.453.

[104] *Od.* 21.153–154, 21.170–171. Cf. Onians (1998: 94): "The *thymos* is constantly spoken of as feeling and thinking, as active in the lungs (*phrenes*) or chest (*stêthos*) of the living person, and as departing at death, but it is not spoken of in connection with the succeeding state." On *thymos* as the constituent matter of consciousness in the *phrenes*, cf. 23–24, 30–31, 40, as "the seat of emotions," 44–61.

[105] *Il.* 13.671–672, 16.606–607, 856–857 = 20.362–363; *Od.* 11. 219–22.

Psychê represents the undefinable subtle reality that *substitutes* the physical body after death.[106] A *substitution* looks like a shadow,[107] or dream,[108] or a copy,[109] which no longer has material consistency, but maintains that *form* of living body that guarantees the identity of individuals.[110] Odysseus in Hades recognizes the appearance of all his old friends, namely, of the soothsayer Tiresias, of his mother, and of many other personages he encounters. The immaterial form of the soul is demonstrated by the fact that Odysseus attempts three times to embrace the soul of his mother, but three times her soul flies away from his hands, "like a shadow or dream."[111]

We note that souls in Hades retain the same character in their behavior as they had while living.[112] In Homer, both the emotive and the cognitive spheres of living beings are dominated by *thymos*, which has a precise location in the living body[113] and, belonging to the sphere of corporeality, is never referred to in relation to the souls in Hades. Nonetheless, at a cognitive level, these souls assume the capacity of communicating "true things" to Odysseus[114] by drinking the smoking black blood of the victims he sacrificed during the rite of entry,[115] in so doing, entering again into contact with a hot and fundamental element of the living body, considered to be the site, in fact, of the *thymos*.

Onians notes that the *psychê* "is *in* the person, but it is not spoken of as something which is found in the lungs and chest, nor which thinks or feels as long as the person is alive. It seems rather to be a 'vital principle' or soul, not involved with the state of ordinary consciousness, an entity that persists, still devoid of thusly made consciousness, in the dwelling of Hades, where it identifies with the *eidôlon*, the visible but impalpable semblance of the deceased."[116]

In this archaic vision, it seems therefore that the *psychê* is recognizable as a state of being that immaterially preserves the form (*eidos*) of each living person, even in Hades; it appears, that is, to represent its permanent principle of *identity*.[117] The

[106] On the idea of *psychê* as *substitute* in Hades of the living body, cf. Meyer (2008: 12–15).

[107] *Od.* 10. 495, 11.207.

[108] *Od.* 11.207 and 222.

[109] Odysseus's friend Elpenor appears to him as an *eidôlon* in *Od.* 11.83.

[110] Meyer (2008: 12–13).

[111] *Od.* 11.204–208.

[112] The behavior of Anticlea, mother of Odysseus, was exemplary when she addresses her son whom she has just met in Hades, as observed by the Peripatetic philosopher Praxiphanes (cf. comment to fr 25 by Praxiphanes in Matelli (2012b: 296–297)).

[113] The *thymos* is active in the lungs and in the chest, as well as in the blood: see Onians (1998: 23–50, 94).

[114] *Od.* 10.536–537 = 11.49–50 and 11.88–89 (Odysseus does not permit anyone to drink the sacrificial blood before interrogating Tiresias), 11.96–99, 146–149 (in Hades, Tiresias drinks the sacrificial blood in order to proclaim the "truth" to Odysseus; the other shades as well can tell the truth only after drinking this blood), 11.142, 153 (Anticlea drinks "smoking black blood," before speaking to her son), 11.225–234 (other women in Hades).

[115] *Od.* 11.34–36, cf. Jahn (1987: 36).

[116] Onians (1998: 93–94).

[117] Jahn (1987: 35–37). By identity, I do not mean *consciousness*, which underlies a different debate, in relation to which, see note 4 above.

commonplace according to which the souls of the Homeric heroes in Hades are "larvae"[118] therefore appears to me entirely inexact (I prefer to attribute such definition rather to Orphic beliefs that admit to metempsychosis. Let us recall that in the Hellenistic age, the *psychê* is the moth, the nocturnal butterfly[119]), while it is certainly correct to understand the *psychê* as a *double* of the individual,[120] endowed with the same form as the physical body,[121] even if devoid of matter, like a shadow.[122] The *psychai* encountered by Odysseus in Hades in Book XI of *The Odyssey* always have an easily recognizable appearance, endowed therefore with all the limbs that formed the physical body, even if they are not named. It is striking that in Hades there remains a *bodily form*, even though without *physical matter* (*hylê*). *Psychê* (as *spiritus*) etymologically expresses the breath: it is therefore a spiritual entity, but not in the psychic sense.

7 The Organism in Aristotle

With a bold leap we now go directly to Aristotle, aware of passing over the evolutions of the Homeric beginnings in the successive epic, monodic lyric, and choral poetry, in the Orphic reflection, in the philosophical thinking from the pre-Socratics to Plato, and in the great tragic and comic poets of the fifth century BC. All these authors came up with concepts (sometimes even opposed to one another) which are at the basis of all Western thought articulated on the organism of the living being, and Aristotle is indebted to them.[123] The soul-body dualism, imposed by the Orphic religion, re-elaborated by Pythagoreans and formalized definitively by Plato, was first assimilated by Aristotle in his youthful works, but then superseded in his mature works within a *holistic* vision, in which the natural body has a fundamental philosophical centrality. Ancient medicine only gradually acquired notions adequate for understanding the complexity of the living organism and, in the *Corpus Hippocraticum*, in fact, there still is lacking an *organic* conception of the body,

[118] Mondin (2001: 25).

[119] I refer the reader to Matelli (2004: 318–19 note 55), for the observation that the term *psychê* is used for the first time to indicate the butterfly in the Peripatetic context in the fourth to third centuries BC by Aristotle and Theophrastus, with explicit reference to the metamorphosing insect.

[120] I attribute to the term *double* a value different from Rohde's animist value (1921), criticized by Jaeger (1953: 92–106), who, however, maintains this meaning. See note 101 above.

[121] Jahn (1987: 35–36).

[122] In Calzecchi Onesti's Italian translation of *Od.* 10.494–495 "a lui solo concesse Persefone d'aver mente saggia da morto, gli altri come ombre *vane* svolazzano" ("him alone [i.e., Tiresias] did Persephone permit to have a wise mind in death, the others flutter about as *vain* shades,") the adjective *vain*, absent in the Greek text, is added arbitrarily. The verb *aissô* contains the idea of momentum, cf. Jahn (1987: 36).

[123] The reader is referred to the syntheses of studies cited in note 4 above, to which Sichkaryk (2011: 86–99) can be added.

"which is not perceived and described as an ensemble of reciprocally interrelated organs and functions, but rather as a 'hollow recipient' in which fluids flow and combine."[124] Despite the fact that in archaic times the functioning of the human body was already considered a useful model, especially for the good or poor health of the *polis*,[125] it is only with Aristotle, son of the physician Nicomachus, and with the successive Hippocratic medicians[126] that the living body begins to be understood as an *organism*, and in this more complex meaning, to be used as a model for other types of human organizations.

In nearly all his works, Aristotle speaks of *bodies*, studying them in the biological sense above all in the books on *animals*. Nonetheless, here I will be considering only the idea of *sôma*, in the mature—and synthetic—meaning contained in the three books *On the Soul*. First of all, *natural bodies* (both living and inanimate) are considered as the beginnings (*archai*) of all the others, that is, of artificial bodies, as well.[127] In observing the passage of organisms from life to death, Aristotle seems to give continuity to the Homeric observations according to which the disintegration of the elements constituting the physical body subsequent to death is determined by the abandonment of the limbs by the soul (*psychê*).[128] This latter is therefore a unifying principle, precisely an *eidos*, that is, a *form*: a principle that, in Aristotle then, takes on a new value,[129] a harbinger of a *hylomorphic*, and not dualistic, vision of the relationship of the soul with the living organism,[130] even while admitting that at least one of its functions (that of the intellective soul) survives the physical body.[131] Leaving aside the critical investigation of these themes,[132] I shall examine only two steps, which are necessary to highlight the speculative aspects that are binding here:

1. The soul is presented as *substance*, *form*, and *actuality* (namely, *realization*) *of the physical body*:

> *On the soul* 2.1 (412a19–22)
>
> It is therefore necessary that the soul be substance (*ousia*) as *form of the natural body* (*eidos sômatos physikou*) having life in potentiality (*dynamei zôên echontos*). Such substance is *actuality* (*entelecheia*): therefore the soul is the *actuality of such body* (*toioutou entelecheia sômatos*).

---

[124] Caserta (2007: 66).

[125] Caserta (2007, 2012).

[126] Vegetti (1983: 459–469).

[127] *An.* 2.1 412a11–13 (henceforth *An.*).

[128] *An.* A5, 411b7–8: "it seems, rather, that it is the soul that keeps the body united, since, once the soul leaves, the body suddenly dissolves and putrefies."

[129] Busche (2001: 118–119), Centrone (2005): 103–114), Bodson (2003), Bouquiaux (2003), Évrard (2003), Fiasse (2003), Guldentops (2003), Motte (2003), Opsomer (2003a, b, c, d), Seron (2003), Rutten (2003a, b), Stevens (2003a, b), Vancamp (2003), Fronterotta and Leszl (2005: 180–185).

[130] Meixner, Newen (2003: 52–56).

[131] *An.* 3.5 430a10–25.

[132] For a synthesis, Grasso & Zanatta (2005: 244–263).

2. There is an analogy in the organic structure of all natural bodies, whose unity is given by the soul, which, in giving form to material, makes the life of the body possible:

On the Soul 2.1 (412a27–b9)

Therefore the soul is the first actuality (*entelecheia hê prôtê*)[133] of a natural body that has life in potentiality: such is the body equipped with organs (*organikon*). Organs (*organa*) are also the parts of plants, but extremely simple. Thus, the leaf covers the pericarp and the pericarp the fruit. The roots, then, are analogous to the mouth, since they both draw nourishment. If, therefore, a common definition of each kind of soul is to be proposed, it will be the *first actuality* of a natural body equipped with organs. Therefore, one should not look into whether the soul and the body are one, just as one does not do for the wax and the imprint, nor, in general, for the matter of each thing and the form from which such matter is taken: the *one* and *being*, in fact, are referred to using multiple meanings, but the fundamental one is the *actuality*.

The soul, according to Aristotle—besides appetite and the locomotive powers—consists of three fundamental parts that can explain the forms of all living beings: the *nutritive*[134] faculty, the *sensitive*[135] faculty, and the *intellective*[136] faculty. All plants have, at least, the first faculty; animals have the first two; and humans, all three.[137] The parts of the soul have an intrinsic unity among one another, which not only Aristotle,[138] but also the latest researches, recognize as being *systemic*[139] in nature. But it is, above all, the soul in itself, *animating* the body, that creates a system for all effects, of which the *vegetative life* and the *sensitive life*[140] represent the first *emergent effects*; the *intellective* capacity (*nous*) is, on the other hand, the *emergent effect* typical of the human system and the only to survive death.[141] The soul as a whole is therefore the full, complete reality of the body, that is its *actuality*. But this also requires, in turn, an *acting cause*, which metaphysical thought recognizes in the *active intellect*, the *nous poietikos*, to which the sphere of human intellect is directly connected.[142]

This proportional *balance* between the parts typical of a flourishing organism is functional to offering a model of good health to the life of a state according to

[133] "Ossia il principio formale, la prima e fondamentale determinazione del vivente, condizione di tutte le sue funzioni vitali, cognitive e operative," ("That is, the formal principle, the first and fundamental determination of the living, the condition of all its vital, cognitive, and operative functions,") Movia (2001: 266 note 12); cf. also Busche (2001: 126–131).

[134] *An.* 2.4 (415a14–416b31).

[135] *An.* 2.5 (416b32–418a25).

[136] *An.* 3.4 (429a10–430a9).

[137] *An.* 2.1 (414a29–31).

[138] *An.* 1.5 (411a5–14).

[139] Bastian (2010); Busche (2001: 2 n. 9) intends to overcome the negative judgment of Gigon (1986: 158) according to which the idea of *system* cannot be attributed to Aristotle's philosophical *corpus*.

[140] *An.* 2.7 (418a26–424b18).

[141] *An.* 1.4 (408b19–30), 2.1 (413b24–26), 3.5 (430a10–25), cf. *Metaph.* 12.3 (1070a24–26).

[142] *An.* 3.5 (430a10–25). In view of the exegetical complexities of this section, I refer the reader to the interpretation of Grasso & Zanatta (2005: 244–263).

Aristotle, who takes up again a branch of the Pythagorean[143] tradition, also exploited by Plato[144]:

> *Politics* 5.3 (1302b33–42)
>
> Revolutions in the constitutions also take place on account of disproportionate (*para to analogon*) growth; for just as the body[8] is composed of parts, and needs to grow proportionately in order that its symmetry may remain, and if it does not it is spoiled, when the foot is four cubits long and the rest of the body two spans, and sometimes it might even change into the shape *(morphê)* of another animal if it increased disproportionately not only in size but also in quality, so also a state is composed of parts, one of which often grows without its being noticed, as for example the number of the poor in democracies and constitutional states.[145]

Analogously, he uses analogies with the physiology of the human body in order to explain the creative phenomenon of poetic art: referring to the literary genre of tragedy, for example, he observes that, after the many changes at the moment of growth, this "ceased changing when it *reached its own nature*."[146] The body of a poetic work, "*like that of an animal, and of anything composed of parts* (*kai zôon kai hapan pragma*)," must be equipped with a "nonrandom" magnitude and "have its parts well arranged in a precise order (*tetagmena*)."[147] In the *Rhetoric*, then, through the metaphor of the physical body, he presents the argument as the *sôma* of the *entimema*[148] and, by means of similitudes with the limbs of the human body, highlights the necessity of a proportional order of the parts (*taxis*) with respect to the whole.[149]

It is interesting to note how the analogy between the structural proportions and the organization of the natural body with other structurally complex realities was in any case already present for some time in the rhetorical genre of the *paradigmatic fable*.[150]

Dio Chrysostom, in a discourse to the inhabitants of Tarsus, made use of a fable attributed to Aesop to explain the specificity of the functions and the necessary difference and hierarchy of the social parts:

> Dio Chrysostom, *Discourses* 33.16
>
> Something must have happened to you like what Aesop says happened to the eyes. They believed themselves to be the most important organs of the body, and yet observed that it

[143] In his *On Nature*, 24 DK B4, Alcmaeon speaks of the state of physical *health* (*hygieia*), sustaining that it consists in an equilibrium (*isonomia*) of opposing (as in humid/dry, cold/hot, sweet/bitter, etc.) forces (*dynameis*), while the state of illness (*nosos*) comes about through the prevalence of just one of these forces over the others, a situation he refers to using the term *monarchy*: note the lexicon derived from the political sphere being used to qualify the prosperity or pathology of the physical body. Cf. Caserta (2007: 70).

[144] Plat. *Tim.* 15 42e ss.

[145] I am following Rackham's translation of Aristotle's *Politics*.

[146] Arist. *Poet.* 4 (1449a9–15). Cf. Plat. *Phaedr.* 264c.

[147] *Poet.* 7 1450b32–1451b6. Cf. also the analogy of *Poet.* 23 1459a17–24. Matelli (2012b: 419–450).

[148] *Rhet.* 1.1 1354a11–16.

[149] *Rhet.* 3.14 1415b6–8, 3.19 1496b19–23. Matelli (2012a: 725–754).

[150] Adinolfi (1963: 333–342).

was the mouth that got the benefit of most things and in particular of honey, the sweetest thing of all. So they were angry and even found fault with their owner. But when he placed in them a bit of honey, they smarted and wept and thought it was a stinging, unpleasant substance (translated by Cohoon and Crosby 1951).

The order (even hierarchical) of the social parts of the organism of the state is clarified using the same paragon by Dionysus of Halicarnassus[151]; Livy refers to the apologue with which Menenius Agrippa resolved a social discord in 493 BC, reminding the rebels of the connection between the well-being of the individual members and the entire organism.[152] In ethical discourses, Cicero, Seneca, and Epictetus reaffirm the necessity of overcoming individual egoism and of understanding the necessary connection between the well-being of the individual organ and that of the entire organism of which that is part:

Cicero, *On Duties* 3.5.22

If each member of our body were to imagine that it could be healthy and strong by attracting to itself the health and vigor of the neighboring limb, then necessarily the entire organism would weaken and perish; similarly, if each of us were to appropriate the assets of others, subtracting from each one as much as possible for his own advantage, necessarily human society would fall into ruin.

Seneca, *On Anger* 2.31.7.

What would happen if the hands should desire to harm the feet, or the eyes <to harm> the hands? As all the members of the body are in harmony one with another because it is to be the advantage of the whole that the individual members be unharmed, so mankind should spare the individual man, because all are born for a life of fellowship, and society can be kept unharmed only by the mutual protection and love of its parts (transl. by Basore 1958).

Seneca, *Moral Letters to Lucilius* 95.52

All that you see, which includes the divine and human spheres, forms a unity: we are members of one great body. Nature has drawn us toward a life bound by ties of kinship, generating us from the same principles and to tend toward the same ends. It has infused within us a mutual love and has made us prone to solidarity.

Epictetus, *Discourses* 2.10.4

What then is the profession of the citizen? Not having anything for one's own utility, nor to organize anything as though against unity, but to act like the foot or the hand which, if they were to have the possibility of reasoning or understanding the constitution of nature, would never exert choice or desire in any other way, except in reference to the whole.

The historian Josephus utilizes the same metaphor to explain the spreading of a disgruntled political movement in Judea:

Josephus, *The Jewish War* 4.7.2 (406)

Moreover, in other regions of Judea, the bands of brigands entered into action, which previously had not moved, as happens in the body that falls ill, and all the other parts feel it.

This is a topos that, in great chronological proximity, was also chosen by St. Paul to communicate to the Greeks in Corinth the novelty of Christianity, which made them *ontologically* "parts" of the *risen body* of Christ:

[151] *Roman Antiquities* 6.83.2.

[152] *History of Rome* 2.32.7–12.

St. Paul, 1 *Cor.* 12, 12–26.

As the body, in fact, though being one, has many members and all the members, even though being many, are only one body, so also is Christ. And in reality, we have all been baptized in one single spirit to form one single body, Jews or Greeks, slaves or freemen; and we have all quenched our thirst from one single spirit. Now, the body turns out to be not one single member, but of many members. If the foot were to say: "Because I am not a hand, I do not belong to the body," it would not, for this, cease being a part of the body. And if the ear were to say: "Since I am not an eye, I do not belong to the body," it would not, for this, cease being a part of the body. If the body were entirely eye, where would the hearing be? If it were entirely hearing, where would the sense of smell be? Now, instead, God has arranged the members in a distinct way in the body, as He has willed. If, then, everything were only one member, where would the body be? Instead, the members are many, but the body is only one. The eye cannot say to the hand: "I do not need you"; nor the head to the feet: "I do not need you." Indeed, those members of the body that seem weakest are the most necessary; and those parts of the body that we hold to be less honorable, we surround with greater respect, and those that are indecorous are treated with greater decency, while those that are decent do not have need of this. My God has composed the body, conferring greater honor upon that which was therein lacking, in order that there would not be disunity within the body, but that instead the various members would take care of one another. Therefore if one member suffers, all the members suffer together; and if one member is honored, all the members rejoice with that one. Now, you are the body of Christ and his members, each one for his own part.

8 Conclusions

The line of thought followed up to this point has demonstrated how the organicist model has quite ancient roots. Already in Homer's poems, the idea of a *form* (called *soul*) is present as *principle of unity and organization of the members*, contrasting the forces which, in the biological body, would tend otherwise toward disintegration. Aristotle introduces, furthermore, the idea that the soul, in order to perform its function, should in turn be associated with an *acting principal*, intellective, and recognized in the metaphysical *nous*.

Plato seems to be the first to use the term "system" (*systêma*) in reference to the political and institutional structure of the state[153] and to the organization of the intervals (*diastêmata*) that form musical harmonies,[154] as well as for a synonym for *systasis* to express the unity of "numeric systems" made possible by a unifying "bond" (*desmos*).[155] Aristotle, as well, would use the term *systêma*[156] various times, without, however, dedicating, at least explicitly, specific analysis to the theme; in reality, it is the treatment of *sôma*, above all in the treatise *On the Soul*, that contains the theoretical foundations that we find again in

[153] *Laws* 686b7.

[154] *Philebus* 17d2.

[155] *Epinomis* 991e2.

[156] Aristotle, *Nicomachean Ethics* 1168b.32, *On the Generation of Animals* 740a.20, 752a.7, 758b.3, *On the Heavens* 391b.9, *Poetics* 1456a.12.

the successive theory of systems. The living body is, in fact, an organized unity composed of parts, each one with a specific function, connected by internal constraints and by hierarchies, sensitive to the external environment and to the passage of time, with an emergent effect with respect to the properties of the individual parts. By guaranteeing unity, the functional organization of the parts gives identity to the living body, thanks to rules that bind its behavior, both from above, and beginning from the alterations in its basic constituents.[157] *Psychê* is the *efficient* and *formal cause* of this extraordinary organization; however, this emergent property is only possible because it is itself a *system*[158] that receives *cause* and *form* from another, that is, from a superior and active intelligence (the *nous poietikos*).

I shall conclude with a text that opened a new perspective in the first century AD: in the first pericope we find synthesized ideas that are already present in classical Greek thinking, but, by means of the introduction—in the second part—of the theological datum of the *incarnation* of God and the *resurrection of the body* of Christ, the new Adam, St. Paul elaborates an idea of *body* that is absolutely original, and which, in resuming the key ideas established by Aristotle (even if with a different terminology), is capable of overcoming the philosopher's metaphysical aporia[159]:

St. Paul, *1 Cor.* 15, 44–45

If there is an *animated body* (*sôma psychichon*), there is also a *spiritual body* (*sôma pneumatikon*), because it is written that the *first man, Adam*, became *a living being* (*ho prôtos anthrôpos Adam eis psychên zôsan*), but the *last Adam* became a *life-giving spirit* (*ho eschatos Adam eis pneuma zôopoioun*).[160]

Bibliography

Adinolfi, M. (1963). *Le metafore greco-romane della testa e del corpo e il corpo mistico di Cristo*. In *Studiorum Paulinorum Congressus internationalis catholicus 1961, simul Secundus Congressus internationalis catholicus de Re Biblica, completo xix saeculo post S. Pauli in Urbem adventum* (Vol. II, pp. 333–342). Roma, Pontificio Istituto Biblico (Analecta Biblica 18).

Aristoteles. (1932). Aristotle's *Politics*, with an English translation by H. Rackham. Cambridge, London: Harvard University Press.

Aristoteles. (1961). Aristoteles, *De Anima*, Edited with Introduction and Commentary by W.D. Ross. Oxford: Clarendon Press.

Bartolomei, M. C. (1984). *Ellenizzazione del Cristianesimo: linee di critica filosofica e teologica per una interpretazione del problema storico*. L'Aquila-Roma: Japadre.

Basore, J. W. (1958). Seneca, *On Anger*. In Seneca Moral Essays, with an English translation. (Vol. I, pp. 106–355). London: William Heinemann Ltd, Cambridge: Harvard University Press.

157 The reader is referred to the contributions in this volume by Urbani Ulivi and Giuliani.

158 See note 139 above.

159 See Appendix IV on "Ancient Jewish conceptions of the mind or 'soul', the 'spirit', 'the holy spirit', the body and the divinity of Christ" in Onians (1998: 480–505); Grasso & Zanatta (2005: 244–263) and Busche (2001: 100–146).

160 I propose a literal translation of this holy text.

Bastian, T. (2010). *Die Seele als System*. Göttingen: Vandenhoeck & Ruprecht.

Beekes, R. (2010). *Etymological dictionary of Greek* (Vol. 1–2). Leiden-Boston: Brill.

Bertalanffy, L. (1969). *General system theory: Foundations, development, applications*. New York: George Braziller.

Bertolaso, M. (2013). Sull'irriducibilità della prospettiva sistemica in biologia. Un'analisi delle convergenze dei modelli interpretativi del cancro. In L. Urbani Ulivi (Ed.). *Strutture di mondo. Il pensiero sistemico come specchio di una realtà complessa* (Vol. II, pp. 143–169). Bologna: Il Mulino.

Bodson, L. (2003). *§ 6. Aristote,* Génération des animaux, Histoire des animaux, Marche des Animaux, Mauvement des animaux. In A. Motte, C. Rutten, & P. Somville (Eds.), *Philosophie de la forme: "eidos", "idea", "morphè" dans la philosophie grecque des origines à Aristote* (pp. 391–433).

Böhme, J. (1929). *Die Seele und das Ich im homerischen Epos*. Leipzig-Berlin: Teubner.

Bouquiaux, L. (2003). *§ 1. Aristote,* Du ciel. In A. Motte, C. Rutten, & P. Somville (Eds.), *Philosophie de la forme: "eidos", "idea", "morphè" dans la philosophie grecque des origines à Aristote* (pp. 341–359).

Busche, H. (2001). *Die Seele als System: Aristoteles' Wissenschaft von der Psyche*. Hamburg: Meiner.

Bellia, G. (2011). *L'ellenizzazione del cristianesimo: un percorso storico-teologico. Introduzione*, in Bellia, G. and Garibba, D. (Eds.), *L'ellenizzazione del cristianesimo dal I al II secolo*. Atti del XIII Convegno di Studi Neotestamentari (pp. 5–13). Bologna: EDB.

Bellia, G. and Garibba, D. Eds (2011). *L'ellenizzazione del cristianesimo dal I al II secolo*. Atti del XIII Convegno di Studi Neotestamentari (Ariccia, 10–12 settembre 2009), Bologna: EDB.

Calzecchi Onesti, R. (Ed.). (1950). Omero, *Iliade*. Torino: Giulio Einaudi Editore.

Calzecchi Onesti, R. (Ed.). (1963). Omero, *Odissea*. Torino: Giulio Einaudi Editore.

Caserta, C. (2007). *Normale e patologico nel corpo e nella polis: "isonomia" e armonia fra VI e V secolo*. In G. Daverio Rocchi (Ed.), *Tra concordia e pace: parole e valori della Grecia antica*, Giornata di studio, Milano, 21 ottobre 2005 (pp. 65–87). Milano: Cisalpino (Quaderni di Acme 92).

Caserta, C. (2012). Πόλεων καὶ ἰδιωτῶν κράτιστοι: "corpo politico" e salvezza della città nell'ideologia democratica. In S. Cataldi, E. Bianco, & G. Cuniberti (Eds.), *Salvare le "poleis", costruire la concordia, progettare la pace* (pp. 365–383). Alessandria: Edizioni dell'Orso.

Centrone, B. (2005). *L'*eidos *come* holon *in Platone e i suoi riflessi in* Aristotele. In Francesco Fronterotta, Walter Leszl (a cura di), *Eidos - Idea. Platone, Aristotele e la tradizione platonica* (pp. 103–114). Sankt Augustin: Academia Verlag (International Plato Studies).

Chantraine, P. (2009). *Dictionnaire étymologique de la langue grecque.* Histoire des mots, nouvelle edition, Bonchamps-lès-Laval, Klincksieck.

Cicero. (1971). Marcus Tullius Cicero, *De officiis*, quartum recognovit C. Atzert, Leipzig, B.G. Teubner.

Cohoon, J. W., Crosby, H. L. (1951). Dio Chrysostom, *Discourses* (Vol. III, pp. 31–36). Cambridge: Harvard University Press, London: William Heinemann Ltd.

Del Giudice, E. (2010). Una via quantistica alla teoria dei sistemi. In L. Urbani Ulivi (Ed.), *Strutture di mondo. Il pensiero sistemico come specchio di una realtà complessa* (Vol. I, pp. 47–70). Bologna: Il Mulino.

Dio Chrysostom. (1893). In J. de Arnim (Ed.), *Dionis Prusaensis quem vocant Chrysostomum quae exstant omnia* (Vol. I–II). Berlin: Weidmann.

Epictetus. (1916). In Schenkl (Ed.), *Epicteti Dissertationes ab Arriano digestae*. Leipzig: B. G. Teubner.

Évrard, É. (2003). *§ 5. Aristote,* Génération et Corruption. In A. Motte, C. Rutten, & P. Somville (Eds.), *Philosophie de la forme: "eidos", "idea", "morphè" dans la philosophie grecque des origines à Aristote* (pp. 383–389).

Fiasse, G. (2003). *§ 12. Aristote,* Étiques. In A. Motte, C. Rutten, & P. Somville (Eds.), *Philosophie de la forme: "eidos", "idea", "morphè" dans la philosophie grecque des origines à Aristote* (pp. 507–537).

Freedman, D. N. (Ed.) (1992). *The Anchor Bible dictionary* (Vol. I, pp. 767–772). New York: Doubleday, *s.v.* "body".
Fronterotta, F., & Leszl, W. G. (Eds.). (2005). *Eidos - Idea: Platone, Aristotele e la tradizione platonica*. Sankt Augustin: Academia Verlag (International Plato Studies 21).
Gigon, O. (1986). Physik und Metaphysik in Aristoteles' PERI PSYCHES. In *Energeia. Études aristotéliciennes offertes a Mgr. Antonio Iannone*. Paris: J. Vrin.
Grasso, R., & Zanatta, M. (2005). *La forma del corpo vivente. Studio sul De Anima di Aristotele*. Milano: Edizioni Unicopli.
Guldentops, L. (2003). *§ 11. Aristote,* Politiques. In A. Motte, C. Rutten, & P. Somville (Eds.), *Philosophie de la forme: "eidos", "idea", "morphè" dans la philosophie grecque des origines à Aristote* (pp. 497–505).
Hirzel, R. (1914). *Die Person. Begriff und Name derselben im Altertum* (p. 10). Sitzungen der Königlich Bayerischen Akademie der Wissenschaften.
Homer. (1917^2–1919^2). *Homeri Opera*, recognoverunt brevique adnotatione critica instruxit Th.W. Allen, III–IV (*Odyssea*), Oxonii, e Typographeo Clarendoniano.
Homer. (1920^3). *Homeri Opera*, recognoverunt brevique adnotatione critica instruxerunt D.B. Monro et Th.W. Allen, I–II (*Iliades*), Oxonii, e Typographeo Clarendoniano.
Jaeger, W. (1953). *Die Theologie der frühen griechischen Denker*. Stuttgart: W. Kohlhammer (Ital. trans. *La teologia dei primi pensatori greci*, Firenze, La Nuova Italia, 1961).
Jossa, G. (2011). *Giudaismo ellenistico e cristianesimo ellenistico*. In Bellia, G. and Garibba, D. (Eds.), *L'ellenizzazione del cristianesimo dal I al II secolo*. Atti del XIII Convegno di Studi Neotestamentari (pp. 15–26). Bologna: EDB.
Jahn, T. (1987). *Zum Wortfeld 'Seele-Geist' in der Sprache Homers*. München: C.H.Beck.
Lehrs, K. (1964). *De Aristarchi studiis homericis*. Leipzig: Gebruder Bornträger, 1882^3 = Hildesheim, G. Olms.
Matelli, E. (2004). Grammata didaskein a Rodi nel III-II sec. a.C.. In *La cultura ellenistica. L'opera letteraria e l'esegesi antica*. Atti del Convegno COFIN 2001 (pp. 307–325). Roma, Edizioni Quasar.
Matelli, E. (2011). In che senso si possa parlare di 'sistema poetico' secondo Aristotele. *Rivista di Filosofia Neo-scolastica, 103*, 419–450.
Matelli, E. (2012a). Perché sia necessario parlare di retorica in senso sistemico. *Rivista di Filosofia Neo-scolastica, 104*, 725–754.
Matelli, E (2012b). Prassifane. Testimonianze e frammenti. Filosofia e grammatica in età ellenistica. Milano: Vita e Pensiero (Pubblicazioni del Centro di Ricerche di Metafisica. Temi Metafisici e Problemi del Pensiero Antico).
Meixner, U., & Newen, A. (Eds.). (2003). *Seele, Denken, Bewusstsein. Zur Gesschichte der Philosophie des Geistes*. Berlin-New York: Walter De Gruyter.
Meyer, M. F. (2008). Der Wandel des Psyche-Begriffs im frühgriechischen Denken von Homer bis Heraklit. *Archiv für Begriffsgeschichte, 50*, 9–28.
Minati, G. (2013). Note di sintesi: novità, contributi, prospettive di ricerca dell'approccio sistemico. In L. Urbani Ulivi (Ed.), *Strutture di mondo. Il pensiero sistemico come specchio di una realtà complessa* (Vol. II, pp. 315–336). Bologna: Il Mulino.
Morgan, G. (1986). *Images of organization*. Beverly Hills-Newbury Park, Cal.-London-New Dehli: Sage (Ital. trans. *Images. Le metafore dell'organizzazione*, Milano, Franco Angeli, 1989).
Motte, A. (2003). *§ 13. Aristote,* Rhétorique. In A. Motte, C. Rutten, & P. Somville (Eds.), *Philosophie de la forme: "eidos", "idea", "morphè" dans la philosophie grecque des origines à Aristote* (pp. 539–549).
Motte, A., Rutten, C., & Somville, P. (Eds.). (2003). *Philosophie de la forme: "eidos", "idea", "morphè" dans la philosophie grecque des origines à Aristote*, Actes du colloque interuniversitaire de Liège, 29 et 30 mars 2001, travaux du Centre d'études aristotéliciennes de l'Université de Liège, Louvain-la-Neuve-Paris, Peeters.
Movia, G. (Ed.). (2001). *Aristotele, L'anima. Testo greco a fronte*. Milano: Bompiani.

Matelli, E (2012b). *Prassifane. Testimonianze e frammenti. Filosofia e grammatica in età ellenistica.* Milano: Vita e Pensiero (Pubblicazioni del Centro di Ricerche di Metafisica. Temi Metafisici e Problemi del Pensiero Antico).

Mondin, B. (2001). *Storia dell'antropologia filosofica. Dalle Origini fino a Vico.* (Vol. I). Bologna: Edizioni Studio Domenicano.

Onians, R. B. (1998). *The Origins of European Thought about the Body, the Mind, the Soul, the World, Time and Fate.* Cambridge: Cambridge University Press (Ital. trans *Le origini del pensiero europeo. Intorno al corpo, la mente, l'anima, il mondo, il tempo e il destino: nuove interpretazioni di materiali greci e romani, di altre testimonianze e di alcune fondamentali concezioni ebraiche e cristiane*, Milano, Adelphi, 2011[2]).

Opsomer, J. (2003a). *§ 3. Aristote,* Du Ciel. In A. Motte, C. Rutten, & P. Somville (Eds.), *Philosophie de la forme: "eidos", "idea", "morphè" dans la philosophie grecque des origines à Aristote* (pp. 361–370).

Opsomer, J. (2003b). *§ 4. Aristote,* Météorologique. In A. Motte, C. Rutten, & P. Somville (Eds.), *Philosophie de la forme: "eidos", "idea", "morphè" dans la philosophie grecque des origines à Aristote* (pp. 371–381).

Opsomer, J. (2003c). *§ 8. Aristote,* De l'Âme. In A. Motte, C. Rutten, & P. Somville (Eds.), *Philosophie de la forme: "eidos", "idea", "morphè" dans la philosophie grecque des origines à Aristote* (pp. 441–458).

Opsomer, J. (2003d). *§ 9. Aristote,* Parva Naturalia. In A. Motte, C. Rutten, & P. Somville (Eds.), *Philosophie de la forme: "eidos", "idea", "morphè" dans la philosophie grecque des origines à Aristote* (pp. 459–467).

Ostwald, M. (1969). *Nomos and the beginnings of the Athenian democracy* Oxford: Clarendon Press.

Otto, W. (1923). *Die Manen oder von den Urformen des Totenglaubens. Eine Untersuchung zur Religion der Griechen, Römer und Semiten und zum Volksglauben überhaupt.* Berlin: Verlag von Julius Springer.

Pubblicazioni del Centro di Ricerche di Metafisica.

Pitta, A. (2011). *Sistemi argomentativi e topologie ellenistiche nelle lettere paoline.* In Bellia, G. and Garibba, D. (Eds.), *L'ellenizzazione del cristianesimo dal I al II secolo.* Atti del XIII Convegno di Studi Neotestamentari (43-90). Bologna: EDB.

Rohde, E. (1921). *Psyche: Seelencult und Unsterblichkeitsglaube der Griechen.* Tübingen: J.C.B. Mohr.

Rutten, C. (2003a). *§ 15. Aristote,* Fragments. In A. Motte, C. Rutten, & P. Somville (Eds.), *Philosophie de la forme: "eidos", "idea", "morphè" dans la philosophie grecque des origines à Aristote* (pp. 555–574).

Rutten, C. (2003b). *§ 14. Aristote,* Poétique. In A. Motte, C. Rutten, & P. Somville (Eds.), *Philosophie de la forme: "eidos", "idea", "morphè" dans la philosophie grecque des origines à Aristote* (pp. 551–554).

Schmitt, A. (1990). *Selbständigkeit und Abhängigkeit menschlichen Handelns bei Homer: hermeneutische Untersuchungen zur Psychologie Homers.* Stuttgart F. Steiner.

Schütz, H. G., & Wibbing, S. (2010). *melos*, Glied (*s.v.*). In *Theologisches Begriffslexikon zum neuen Testament* (2nd Edition, pp. 1279–1280). Nördlingen: SCM R. Brockhaus.

Schweizer, E. (1981). "*sôma*" (*s.v.*), In F. Montagnini, G. Scarpat, & O. Soffritti (Eds.), *Grande Lessico del Nuovo Testamento* (Vol. 8, pp. 609–790). Brescia: Paideia.

Scott, W. R. (1992). *Organizations, rational, natural and open systems.* Upper Saddle River, New Jersey: Prentice Hall Inc., A Pearson Education Company.

Seron, D. (2003). *§ 1. Aristote,* Organon. In A. Motte, C. Rutten, & P. Somville (Eds.), *Philosophie de la forme: "eidos", "idea", "morphè" dans la philosophie grecque des origines à Aristote* (pp. 333–359).

Sichkaryk, I. (2011). *Corpo come punto focale nell'insegnamento paolino.* Roma: Editrice Pontificia Università Gregoriana (Tesi Gregoriana, Teologia, 185).

Snell, B. (1955). *Die Entdeckung des Geistes. Studien zur Entstehung des europäischen Denkens bei den Griechen*. Hamburg: Classen Verlag.

Stevens, A. (2003a). *§ 10. Aristote,* Métaphisique. In A. Motte, C. Rutten, & P. Somville (Eds.), *Philosophie de la forme: "eidos", "idea", "morphè" dans la philosophie grecque des origines à Aristote* (pp. 469–496).

Stevens, A. (2003b). *§ 16. Philosophie théorique: essai de conclusion sinthétique*. In A. Motte, C. Rutten, & P. Somville (Eds.), *Philosophie de la forme: "eidos", "idea", "morphè" dans la philosophie grecque des origines à Aristote* (pp. 577–582).

Urso, C. (1997). Riferimenti anatomici nell'Iliade di Omero. *Pathologica, 89*(1), 26–30.

Vancamp, B. (2003). *§ 7. Aristote,* Partie des Animaux. In A. Motte, C. Rutten, & P. Somville (Eds.), *Philosophie de la forme: "eidos", "idea", "morphè" dans la philosophie grecque des origines à Aristote* (pp. 435–439).

Vegetti, M. (1983). *Metafora politica e immagine del corpo negli scritti ippocratici*, in *Formes de pensée dans la Collection hippocratique*, In F. Lasserre & P. Mudry (Eds.). Actes du IVe Colloque International Hippocratique Lausanne, 21–26 septembre 1981 (pp. 459–469). Genève: Droz (Publ. Fac. Lettres Univ. de Lausanne XXVI), .

Viagulamuthu, X. P. B. (2002). *Offering our Bodies as a living Sacrifice to God. A Study in Pauline Spirituality Based on Romans 12.1*. Roma: Editrice Pontificia Università Gregoriana (Tesi Gregoriana, Spiritualità, 7).

Villani, G. (2010). La chimica: una scienza della complessità sistemica ante litteram. In L. Urbani Ulivi (Ed.), *Strutture di mondo. Il pensiero sistemico come specchio di una realtà complessa* (Vol. I, pp. 71–89). Bologna: Il Mulino.

Vivante, P. (1955). Sulla designazione del corpo in Omero. *Archivio Glottologico Italiano, 40*, 39–50.

Wibbing, S. (2010). *sôma*, Körper, Leib (*s.v.*). In *Theologisches Begriffslexikon zum neuen Testament* (2nd Edition, pp. 1281–1285). Wuppertal, Brockhaus, Nördlingen: SCM R. Brockhaus.

Zavalloni, R. (1990). Linee di sviluppo nella concezione della "corporeità" umana. *Antonianum, 65*(2–3), 345–382.

Sentences as Systems: The Principle of Compositionality and Its Limits

Aldo Frigerio

Abstract In this chapter, it is argued on two different grounds that sentences in natural languages can be seen as systems. First, beyond their linear order, sentences exhibit a syntactic hierarchical structure. Therefore, they are structured entities. Although this structure is usually interpreted as independent of meaning, many semanticists believe that syntactic structure indicates the order in which the meanings of the parts are combined. Second, although the principle of compositionality—which states that the meaning of a sentence is a function of the meanings of the parts of that sentence—is valid in general for natural languages, this principle has been shown to have many exceptions, where interpretation does not proceed bottom-up but top-down, from the meaning of the whole to the meaning of the parts. For this reason, a radical version of the principle of compositionality is untenable; if the whole depends on its parts and the parts on the whole, then the sentence is a system that cannot be dissected into separate parts without losing something essential.

1 Introduction

A key topic in philosophy of language concerns the conditions of possibility of a natural language—that is, the features a natural language not only possesses but *must* possess if it is to be considered a natural language. Since Frege, scholars have agreed that the semantics of a natural language is necessarily compositional. The principle of compositionality states that the meaning of a sentence is a function of the meanings of the parts of that sentence. Here, the term "function" highlights the

A. Frigerio (✉)
Department of Philosophy, Università Cattolica S. Cuore, Milan, Italy
e-mail: aldo.frigerio@unicatt.it

L. Urbani Ulivi (ed.), *The Systemic Turn in Human and Natural Sciences*,
Contemporary Systems Thinking, https://doi.org/10.1007/978-3-030-00725-6_9

existence of an algorithm by means of which the meaning of the whole can be derived, based on the meanings of the parts and the rules of composition. It seems clear that the principle of compositionality is, at least to some extent, a valid principle. By way of example, consider the text you are now reading. In all probability, you have never previously encountered some of these combinations of words. Nevertheless, you understand these unfamiliar sentences perfectly. If every new sentence were a new meaning unit and not dependent on its parts—that is, if the principle of compositionality were not in force—it would be difficult to explain why such sentences present no problem of interpretation, as the meaning of each new sentence would require a specific learning as any new word does.

However, this is not the case, and the principle of compositionality explains why we can readily understand new sentences: given the meanings of the parts, we are able to construct the meaning of the whole. This is what you are doing while reading this text. Because you know the meanings of the constituent words, you can understand the meaning of these sentences, even though they are new to you. This explains why dictionaries contain lists of words rather than lists of sentences; if sentences were the smallest units of meaning and did not depend on the meanings of their parts, learning a language would involve learning the meanings of sentences rather than a lexicon and grammar. One might conclude, then, that the principle of compositionality implies an atomistic view of meaning—that in order to know the meaning of more complex linguistic units, it suffices to know the meanings of the simplest units. The aim of this essay is to show that this interpretation of the principle of compositionality is incorrect.

The idea that the sentences of natural languages express something more than the sum of their constituents is not new. Since antiquity, many scholars have asserted the semantic indivisibility of sentences. Aristotle, for example, believed that meaningful discourse, λόγοζ, is a unit; if the nouns are isolated from the predicative nexus, the meaning of the discourse is lost. Within the Indian linguistic tradition, Bhartrihari (sixth-century AD) also affirmed the principle of indivisibility of sentences (*sphota*). He claimed that the meaning of a sentence does not correspond to the meaning of the words that form that sentence but is rather a unity captured by an innate intuition of the subject. Here, I contend that the atomistic view of meaning fails for at least two reasons.

1. Even if the principle of compositionality were valid without restriction, it would not follow that a sentence's meaning is the sum of the meanings that constitute it. Sentences have a syntactic structure that differs from their linear order, and that structure affects semantic interpretation. It follows that a sentence's meaning is not reducible to the sum of the meanings of its constituent words.
2. In any case, natural languages are not entirely compositional. The principle of compositionality is restricted by the fact that the correct understanding of sentences often depends on understanding the linguistic context.

2 Sentences are Structured Entities

2.1 Syntactic Structure

It has been demonstrated for a long time by very convincing arguments that the sentences used by speakers of a language have both a linear order, in which words follow each other,[1] and a further level of organization that may differ from this. This second level of organization is not linear but *hierarchical*, as the morphemes and words that form a sentence combine into increasingly larger, nested constituents. For instance, in the sentence "Two brothers of Paul will arrive soon," we can first distinguish two large constituents: the noun phrase (NP) "two brothers of Paul" and the verb phrase (VP) "will arrive soon." These phrases are formed in turn by smaller constituents—for instance, the NP "two brothers of Paul" can be segmented into a determiner ("two") and another constituent ("brothers of Paul"). This latter constituent can be further segmented into a noun ("brothers") and a prepositional phrase ("of Paul"), and so on. This structure is usually represented by means of a tree diagram, as in Fig. 1, where S stands for sentence, N for noun, V for verb, P for preposition, PP for prepositional phrase, and Adv for adverb.

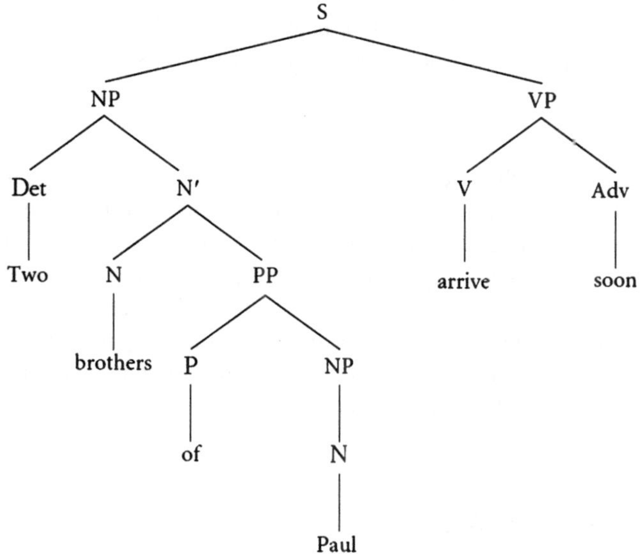

Fig. 1 Syntactic structure of the sentence "Two brothers of Paul will arrive soon"

[1] For written texts, "follow" is intended in a spatial sense; for oral texts, it is intended in a temporal sense.

Such representations presuppose that languages are constituted by a finite number of discrete basic elements (phonemes, morphemes, words). This implies that their division into increasingly smaller constituents must end at a certain point—that is, there are elements that do not contain smaller elements. Additionally, representations of this kind presuppose that the rules that generate sentences are recursive—that is, that elements of a given kind can occur within elements of the same kind. (For instance, the NP "Paul" occurs within the NP "two brothers of Paul.") Recursivity is a property of rules in which the rule can be applied to the result of applying the rule. This property explains the productivity of natural languages. Although they have a finite number of basic elements and a finite number of combinatorial rules, a potentially infinite number of sentences can be generated. The production of this infinite set from a finite initial set can be explained only if the rules that generate larger constituents from smaller ones are recursive.

Rules that enable the generation of sentences presuppose that words are assigned to different categories (noun, verb, preposition, adverb, etc.), and that only some combinations of words from certain categories are allowed. For instance, one may combine an article with a noun but not with a verb (unless the verb is substantivized); while "the dog" is a grammatical phrase, "the goes" is not. Permitted combinations of words from basic categories create larger constituents, which can be combined into still larger constituents, and so on—always on the basis of rules that permit only certain combinations. For instance, the rule governing how a noun phrase and a verb phrase form a sentence can be written as follows:

$$S \rightarrow NP + VP$$

The existence of such structures and rules shows that a sentence cannot be conceived simply as the sum of its constituent words. Beyond their superficial order, sentences have a syntactic structure that links the words into a whole. That structure is governed by rules that specify which relations among the parts are permitted and which are not. This shows that sentences are systems. To account for how language works, it is not sufficient to list its constituent elements (phonemes, morphemes, words) and their meanings; as well as knowing the elements, we must also know how those elements can be structured into wholes.

2.2 A Formal Definition

The thesis that sentences are not mere sequences of words but have a further structure can be more precisely formulated in terms of the kinds of rules and grammars that govern sentence formation. In a regular grammar, the system that generates symbol strings (in natural languages, sentences) has access only to the previous element in the string and can produce only the next element.[2] However, Chomsky (1957) demonstrated that the grammars of natural languages are *not* simple regular grammars, and that more complex grammars—in particular, context-sensitive grammars—must be considered. In such grammars, a new symbol is not generated only on the basis of the previous symbol but is also on the basis of its context.[3] This kind of grammar makes it possible to construe syntactic trees (see Fig. 1), in which the constituents are hierarchically ordered and nested, and the "leaves" of the tree are the words that form the sentence.

2.3 Isomorphism Between Syntax and Semantics

Chomsky characterizes the syntactic structure of sentences as independent of the semantics associated with this structure. However, it is possible to interpret Chomskyan theory in a different way, in which there is instead a more or less perfect correspondence between syntactic and semantic structure. On this view, two words are syntactically connected because of their semantic connection—that is, the connection between their meanings. Syntactic structure would then describe the order in which these meanings must be composed. This isomorphism between syntactic and semantic structure implies at least two prerequisites. First, semantic categories must correspond to syntactic categories—that is, the same *kind* of meaning must correspond to every string belonging to a particular syntactic category, such as NPs. On the other hand, certain semantic operations must correspond to syntactic operations. For example, a given composition of the meanings of V and NP must correspond to the syntactic operation V + NP. Taking account of this semantic-syntactic structure of natural languages, Partee et al. (1990) rephrased the principle of compositionality as follows: the meaning of a complex expression is a function of its constituents and the grammatical rules used to combine them.

[2] This statement simplifies a more complex definition. A regular grammar is a quadruple G = <T, N, E, R>, where T is a set of terminal symbols, N is a set of non-terminal symbols, E is the start symbol, and R is a set of transforming rules $A \rightarrow \beta$, where A is a non-terminal symbol and β is either a terminal symbol or a symbol formed by a terminal symbol and a non-terminal symbol. Given the start symbol, the grammar can generate all the strings of symbols allowed by the rules. Transformation rules can be applied only if the string contains a non-terminal symbol; otherwise, the string is terminal. In natural languages, the start symbol is S (sentence); the non-terminal symbols are constituents larger than words, and the terminal symbols are words. We have already seen an example of a transformation rule ($S \rightarrow NP + VP$).

[3] More precisely, a context-sensitive grammar is a quadruple G = <T, N, E, R>, where T is a set of terminal symbols, N is a set of non-terminal symbols, E is the start symbol, and R is a set of transforming rules $\alpha A \beta \rightarrow \alpha\gamma\beta$, where A is a non-terminal symbol, and α, β, and γ are either terminal symbols or strings of terminal and non-terminal symbols. α and β may be empty; these provide the "context" for A.

3 Limits of the Principle of Compositionality

While formal and logical languages are usually entirely compositional, natural languages are not. In this section, I address some limitations of the principle of compositionality in natural languages, analyzing some phenomena in which the whole conversely determines the meanings of the parts.

3.1 Idioms

Certain natural language expressions, which grammarians refer to as "idioms," do not abide by the principle of compositionality, and their meaning must be specifically learned. Such expressions may be groups of words or entire sentences—for example, "It's raining cats and dogs," "kicked the bucket," or "red herring." Their meaning is not compositional—that is, the meanings of the parts (and the rules of composition) do not suffice to explain the meaning of the whole, as specific knowledge is needed.

3.2 Ambiguity and Polysemy

While idioms are of some relevance, languages are not for the most part idiomatic. However, there are other more pervasive phenomena that limit the principle of compositionality. These phenomena include ambiguity and polysemy. Many words have more than one meaning, and the precise sense in which such words are used is determined by the context, as in the following examples.

"You parked across the street? That's *fine*."
"If you park there, you'll get a *fine*."

The English word "fine" has more than one meaning, and when used in a particular occasion, it expresses only one of these meanings. In most cases, someone who hears or reads a sentence containing an ambiguous word can readily discern the intended meaning from the linguistic context. In this case, it is the context (the whole) that determines the meaning of the word (the part) rather than vice versa. Given that ambiguity and polysemy are widespread phenomena, this represents one important limitation to the principle of compositionality.

3.3 Anaphoric Pronouns

Another pervasive phenomenon concerns the referents of anaphoric pronouns, which are again determined by the immediate context, as in "Ann said to Paul that *he* had to join *her* immediately." Personal pronouns such as "he" or "her" and demonstratives have no reference outside the context in which they are used. They acquire a referent only from the linguistic context (the words that precede and follow) or from the extra-linguistic context (the concrete situation in which they are uttered). The example above is of the first kind; the proper names Ann and Paul appear in the linguistic context, serving as referents for the personal pronouns. In other cases, a pronoun's referent is determined by the extra-linguistic context; for instance, in the case of the sentence "Take *it*," suppose that the speaker is using her finger to indicate a ball. Here, the extra-linguistic context (specifically, the speaker's gesture) determines the pronoun's referent. However, pronouns that acquire their reference from the extra-linguistic context still fulfill the principle of compositionality, as they acquire their referent independently of the linguistic context. This meaning is then integrated with the meanings of other words in the sentence. Rather, the anaphoric pronoun is an example of the whole determining the meaning of a part. Indeed, it is the anaphoric pronoun's presence in one sentence rather than another that provides a certain referent for the pronoun itself. If we change the context, the referent is also changed. In this case, the assignment of the meaning proceeds *top-down* (rather than *bottom-up*).

3.4 Semantic Indeterminacy

Although the phenomenon of semantic indeterminacy bears some resemblance to ambiguity and polysemy, these concepts must be carefully distinguished. As described above, an ambiguous or polysemous word has more than one meaning, and context determines the intended meaning on a given occasion. In the case of semantic indeterminacy, however, there is only one indeterminate meaning, which is determined by the context of use. The following examples (cf. Searle 1980) serve to illustrate this point.

"Ann *cuts* the lawn."
"John *cuts* the cake."

We know that the operation of cutting a lawn is very different from cutting a cake; while the blades of grass are severed using a sickle or a lawnmower, a cake is cut into slices by a knife. It would be surprising if, to cut the lawn, Ann took a knife and performed very long incisions or took a scissors and cut the blades of grass one by one vertically. Similarly, it would be surprising if, to cut the cake, John used a lawnmower. Clearly, the meaning of the verb "to cut" is specified by the context; beyond that, it is vague and indeterminate. In general, it means "to divide something by means of a sharp tool." However, the ways in which the object is divided, the kind of tool used and how it is used are determined by the context and specifically by the object that is cut. For every object, our encyclopedic knowledge suggests the tool to be used to cut it and the ways in which it must be cut, lending the sentence a more determinate meaning. In such cases, the context specifies a meaning that the word would not have in another context.

Semantic indeterminacy is a widespread phenomenon, as for instance in these predications of color noted by Recanati (2004):

"red car"
"red grapefruit"
"red book"

A car can be judged to be red when most parts of its body are red, even though other parts such as wheels, mechanical parts, underside, and interior may not be red. For a grapefruit to be described as red, it must be red internally, although its peel may be another color. A book with a red cover may be described as red although its pages may be another color. In short, the parts of an object that must be red to predicate its redness depend on the object itself and on our encyclopedic knowledge.

As a further example of semantic indeterminacy, consider these sentences.

"Four boys invited three girls"
"Four boys broke three glasses"

These statements can be true in multiple situations. For example, the first sentence is true if four boys together have invited three girls together; or if each of the boys has invited one of the three girls separately; or if each of the boys has invited a different group of three girls; or if two boys together have invited a girl, and other two boys together have invited two other girls together. The sentence may also be true in many other situations, and this also applies to the second sentence. Which of these situations do these sentences describe? The answer is each of them and none in particular—in short, the meaning of these sentences is indeterminate because they may designate any of a number of situations. The first sentence refers to one or more invitations, involving one or more groups of four boys and one or more groups of three girls; only the context can determine the intended meaning.[4] In this case, the semantic indeterminacy is not lexical but concerns how the meanings of different parts of the sentences (noun and verb phrases) are combined. It is not entirely determined how this combination should occur, and many possibilities remain open.

3.5 *Exophoric Pronouns*

As a final example, consider the nature of exophoric pronouns, which refer to objects in the extra-linguistic context. As the contribution of linguistic context to the determination of reference is rarely mentioned, we will analyze this phenomenon in more detail. Suppose that Ann and John are dining at a nice restaurant on the sea. The night is beautiful, and the temperature is perfect. They have just been served and have begun to eat. John says:

---

[4]This is not the usual interpretation of these sentences, which usually is related to the different logical forms underlying the sentence rather than to semantic indeterminacy. In Frigerio (2010), I demonstrated that this solution has many drawbacks and that an interpretation in terms of semantic indeterminacy is preferable.

"It's tasty, isn't it?"

Clearly, John is referring to the food that has just been served. Suppose, however, that he had uttered a different sentence:

"It's beautiful, isn't it?"

It seems likely that John is referring to the restaurant where they are eating or, more generally, to the experience they are sharing. As the situation in which the two sentences are uttered is the same, the difference in the referent of the pronoun "it" must be determined by the predicate of the two sentences. As the predicate "tasty" is usually applied to food, it is plausible to believe that John wishes to refer to the more salient food in that moment: the dish they are tasting. On the other hand, the predicate "beautiful" is usually applied to things that are delightful to look at or, more generally, to pleasant experiences, but not to the flavor of food. The beauty of the restaurant and the setting make reference to the place or the circumstance more probable. Therefore, the predicate can help in determining the referent of the exophoric pronouns.

While it is apparent that an *anaphoric* pronoun acquires its referent from the linguistic context, this is far less obvious in the case of *exophoric* pronouns, which usually refer to salient objects in the utterance context. In some cases, however, there are several candidate referents, and the extra-linguistic context will not suffice. In these cases, the predicate of the sentence can assist identification. Because many predicates are applicable only to certain categories of objects, the predicate is likely to provide information about the category to which the referent belongs while excluding other possible candidates. Moreover, participants in a conversation assume in most cases that what their interlocutors say is true and pertinent to the immediate situation. By "situation" I refer here both to the items of information already exchanged and the extra-linguistic context in which the conversation is taking place, including any events occurring within it. As participants are led to presuppose that their interlocutors speak truly, they are also assumed to predicate true things of the objects to which they refer.[5] If something predicated of a candidate referent of an exophoric pronoun is clearly false (although of the right category), it will probably be discarded, and an alternative candidate will be considered. In John's sentence "It's beautiful, isn't it?," the fact that the restaurant and the panorama are pleasing renders these plausible referents of the pronoun "it."

I conclude that the predicate can play an important role in selecting the referent of an exophoric pronoun because candidates that are from the wrong category, falsely predicated or of no relevance to the predicate are commonly ruled out. This is another case in which the *linguistic* context is crucial for the determination of the meaning of a word although that word refers to something extra-linguistic. Again, the whole determines the meaning of the part rather than the other way around.

[5] Grice (1989) emphasizes that participants presuppose that their interlocutors say true and pertinent things.

4 Concerning Defenses of a Radical Version of the Principle of Compositionality

Some scholars who maintain that the principle of compositionality is valid without exception for natural languages have attempted to account for the phenomena analyzed above. In this section, an evaluation of their proposals reveals that they fail.

4.1 Idioms

The only way to deal with idioms in a radical theory of compositionality of natural languages is to consider them as primitive expressions. This means that "It's raining cats and dogs" or "red herring" should not be seen as formed by words but as expressions that are not reducible to simpler elements. Nevertheless, it is clear that the same linguistic element occurs in both of the following sentences.

"It's raining cats and *dogs*."
"My neighbor's *dogs* barked all night."

Admittedly, in the first sentence, "dogs" has lost any independent meaning; if words are meaningful units, "dogs" is a word in the second sentence but not in the first. However, what determines the word status of "dogs" is the context—the fact that it occurs together with other words and expressions. It is therefore beyond question that the whole here determines the status of the part, as the status of "dogs" depends on the context.

4.2 Ambiguity

When consulting a dictionary, we often come across indexed words such as the following.

(a) Bank$_1$: a broad elevation of the sea floor around which the water is relatively shallow but not a hazard to surface navigation.
(b) Bank$_2$: an institution for receiving, lending, exchanging, and safeguarding money.

Words like this are called *homophones*; bank$_1$ and bank$_2$ are differently indexed as two words that have the same form but different meanings. Words are relationships between a form and a meaning. When a form has two meanings, two relationships between a form and a meaning exist, and, thus, two words.

Advocates of the radical theory of compositionality in natural language affirm that there is compositionality only once the words of the sentences have been identified. We would have two interpretative steps: pre-semantic and semantic. In the pre-semantic step, among other things, the words that occur in the sentence are identified. In the semantic step, their meaning is compositionally derived.

Two remarks are in order here. First, it is simplistic to think of pre-semantic and semantic steps as two successive events. If the pre-semantic step preceded the semantic step, it would difficult to understand the basis for considering $bank_1$ or $bank_2$ as part of the sentence. It seems likely that, in trying to interpret a sentence that contains the form "bank," we examine the context in which that form occurs and then compare the meanings of "$bank_1$" and "$bank_2$" with those of the words in the context before choosing the appropriate word. The decision is taken on the basis of some trial and error mechanism, which we use to compose the different words until we achieve a satisfactory outcome. Only at this point do we identify the right word, and this does not rule out the possibility that our decision can be revised as we move forward in our interpretation of the passage. Secondly, if our decision is determined by the context, then the speaker's intended meaning can be determined only from the context. This limits compositionality because if compositionality arises once the word in question is identified, the decision about which word to opt for is not compositional.

4.3 Anaphoric Pronouns

Similar issues arise in the case of anaphoric pronouns; only once a decision has been reached concerning the referent is it possible to compose the meaning of the words. This decision is usually indicated through co-indexing of the pronouns and their antecedents, as in the following example.

> "Ann_1 said to $Paul_2$ that he_2 had to join her_1 immediately."

The indexes mean that "he" has the same referent of "Paul" and "her" the same referent of "Ann." Again, the decision about indexes is part of the pre-semantic interpretative step. However, there is no doubt that, in deciding on the antecedents of pronouns, it becomes necessary to examine the context in which they occur. If no individual in the extra-linguistic context is of sufficient salience to be the referent of one of the pronouns, it becomes necessary to co-index the pronoun with one of the NPs in the linguistic context. Identifying one of these NPs as the antecedent depends on several factors, including closeness in the superficial order, place in the syntactic hierarchy and semantic plausibility. Examining these briefly, pronouns generally refer to objects that are salient in the context. Anaphoric pronouns in particular refer to salient objects cited in the linguistic context. Such objects are especially salient if they have just been cited or if are the main topic of the discourse. Additionally, higher syntactic positions are more salient because they are the immediate objects of discourse. This explains why closeness in the superficial order and a high place in the hierarchical syntactic order make some potential referents more likely than others. Consider the following example.

> "Yesterday, I met Albert before my session at the gym. On my way out, I met John. *He* had some problems, but now *he* is ok."

Barring the presence of other important contextual factors, the recipient will probably infer that the anaphoric pronoun "he" refers to John rather than to Albert, as the NP "John" is closer in the superficial order, making it more salient than the NP "Albert" at the moment when "he" is uttered.

Now consider another example.

"The father of Robert believes that Ann wants not to see *him*."

Even though "Robert" is closer in the superficial order to "him" than "the father," it is probable that the recipient will interpret the pronoun as referring to Robert's father rather than to Robert. This is because the NP "Robert" is nested in the NP "the father of Robert" and therefore occupies a lower position in the syntactic hierarchy. This was not the case in the previous sentence because "Albert" and "John" occurred in different sentences and were on the same hierarchical level. Now consider the following sentence.

"Paul had a very large money supply, and Charles was in hot water. Nevertheless, *he* did not help *him*."

In this case, the referents of the pronouns are not determined by their superficial order or by their syntactic position but by semantic plausibility. It is natural to think that whoever has a large supply of money can help whoever is in trouble. The use of "Nevertheless" at the beginning of the second sentence signals that things did not happen as one might have expected, making it natural to co-index "he" with "Paul" and "him" with "Charles."

As I have tried to show briefly in these examples, co-indexing is a complex process that depends on a plurality of factors. In any case, syntactic and semantic context plays a crucial role in the use and interpretation of pronouns. In this case, as interpretation can be seen to proceed as a process of trial and error, the decision about the correct co-indexing is non-compositional.

4.4 Semantic Indeterminacy

With regard to the contextual determination of the meaning of words, advocates of the radical interpretation of the principle of compositionality have advanced the following defense. Semantics studies the abstract meaning of words prior to their insertion in a certain context. It is true, they admit, that the meaning of words is determined in the context. In fact, the abstract meaning of words may have determinable variables—parts that remain indeterminate while waiting for contextual determination. However, they argue, the principle of compositionality relates to the abstract meaning of words rather than to their contextual meaning. This abstract meaning is composed with that of other words in the sentence before being contextually determined (cf. Predelli 2005).

However, it is unlikely that the principle of compositionality concerns *only* the abstract meaning of words, and certainly not in the case of pronouns. In trying to interpret a sentence like "She will come tomorrow," we must ascribe the predicate "come" to a specific individual. It seems probable that we do not (or do not only) combine the non-contextual meaning of "she" (which is, more or less, a human female other than the interlocutors) with that of the verb "to come" to derive an abstract meaning that we then try to determine by supplying the referent of "she." As it is the referent of "she" who will come tomorrow, it seems probable that we (also) combine this referent with the meaning of the verb in order to construct the meaning of the sentence as a whole. What happens in the case of pronouns probably also happens in other cases of contextual determination of meaning. For instance, it is unclear why we should deny that the contextual meaning of "to cut" combines with that of other words in the context, such as "hair." In these cases, it is also probable that interpretation of the sentence's meaning involves many steps, proceeding from the meaning of the parts to the meaning of the whole and vice versa. Admittedly, the starting point is the abstract meaning of "cut" and "hair" and the syntactic structure of the sentence, in which the NP "the hair" is the complement of the verb "to cut." It seems probable that the abstract meanings of the two terms are initially composed. However, this result engenders a process of determination. On the one hand, it seems improbable that the NP refers to all existing hair, as a single act of cutting all the world's existing hair conflicts with our encyclopedic knowledge. Rather, it relates to a specific instance of hair (for instance, of a specific person) and, the interpreter will therefore search for something in the context that would restrict the denotation of this NP. On the other hand, our encyclopedic knowledge suggests that the cutting of hair is carried out using certain tools and in a certain way, so restricting and determining the meaning of "to cut." At this point, the meanings of the parts, so determined, are composed again, and the new meaning of the whole sentence is obtained, more determinate than the previous one.

If this is the process through which we interpret expressions such as "to cut the hair," we may conclude that the composition of meaning relates to both abstract and contextual meaning. There may also be intermediate compositions. The composition of abstract meanings may prompt a process of determination of the parts. When the meanings of the parts, so determined, are composed again, this may produce a new, more radical determination of the meaning of the parts, and so on repeatedly until the final result.

5 Conclusion

This chapter defends two opposite theses. First, the centrality of the principle of compositionality has been asserted, as without the bottom-up processes of compositionality, we could not utter new sentences and hope to be understood. On the other hand, radical versions of the principle of compositionality cannot be accepted, and some aspects of natural language can be understood only if the context determines the meanings of the elements in a typically top-down process. At this juncture, a question naturally arises: are these two theses contradictory? While in-depth analysis of this problem is beyond the scope of this essay, we can suggest that there is at least one way of interpreting these theses as non-contradictory. To do so, we must understand the relation between words and context as a virtuous circle. It is because a word, literally and conventionally, has a certain meaning that a certain context is selected—that is, that a certain encyclopedic knowledge is mobilized and the recipient's attention is directed to a certain portion of the world. On the other hand, it is the context so activated that permits the meaning of a word to be determined. The mobilized encyclopedic knowledge and the fact that the meaning of a word must be composed with the meaning of another word triggers the process of determination of the two words' meanings. Because these meanings must be "put together" in a certain context, they must assume a certain determination in order to be so composed. If it is true that, in a sentence, the whole depends on its parts and the parts on the whole, then the sentence is a system that we cannot dissect into its separate parts without losing something essential.

References

Chomsky, N. (1957). *Syntactic structures*. The Hague-Paris: Mouton.
Frigerio, A. (2010). *L'unità dell'enunciato. Indeterminatezza semantica e determinazioni pragmatiche*. Milano: Mimesis.
Grice, P. (1989). *Studies in the ways of words*. Cambridge: Harvard University Press.
Partee, B., ter Meulen, A., & Wall, R. (1990). *Mathematical methods in linguistics*. Dordrecht: Kluwer Academic Press.
Predelli, S. (2005). Painted leaves, context, and semantic analysis. *Linguistics and Philosophy, 28*, 351–374.
Recanati, F. (2004). *Literal meaning*. Cambridge: Cambridge University Press.
Searle, J. (1980). The background of meaning. In J. Searle, F. Kiefer, & M. Bierwisch (Eds.), *Speech acts theory and pragmatics* (pp. 221–232). Dordrecht: Reidel.

Mind and Body. Whose? Philosophy of Mind and the Systemic Approach

Lucia Urbani Ulivi

Abstract The debate in Philosophy of Mind though heated is mostly limited to sterile and formal discussions. In order to overcome such a deadlock, I suggest a new breakthrough exploring three paths: rethinking the status and liability of the fundamental assumptions of the discipline; updating the description of human being in view of recent discoveries in neurosciences; and introducing new comprehension instruments, specifically the concept of "system."

In this essay, I first critically consider the reductionist approach that most philosophers of mind accept without question—together with its derivatives such as materialism, scientism, physicalism, mechanicalism—and I ask: do we have reasons for accepting them? Should we revise them, or abandon them, and why?

Secondly, a new picture of human biology comes into focus from leading neuroscientists: brain is plastic and is reshaped by individual experiences; body is a process whose stability is guaranteed by constraints; there is a strong interconnection among bodily activity, feelings, and mind.

Previous considerations drive us to look at mind and body not as separate entities, but as constituents of the same global entity, the human being.

Finally, the concept of "system" is introduced to suggest a solution for the mind-brain problem: in systemic terms, mind is an emergent phenomenon, while brain is a subsystem with respect to the human being.

"What is the relationship between mind and body?" This question has been answered by all the thinkers who have delineated a philosophical theory of humankind and who, in different ways and at different levels of explanation, have considered the mind, the body, and their relationships as a problem within philosophical anthropology. During the twentieth century, especially in Anglophone philosophy,

L. Urbani Ulivi (✉)
Department of Philosophy, Catholic University of the Sacred Heart, Milan, Italy
e-mail: lucia.ulivi@unicatt.it

L. Urbani Ulivi (ed.), *The Systemic Turn in Human and Natural Sciences*, Contemporary Systems Thinking, https://doi.org/10.1007/978-3-030-00725-6_10

the question freed itself from the vast anthropological landscape containing it up to that point, and it acquired an independent status, narrowing the domain of the problem to just two terms: mind and body, or even more directly, mind and brain. From this autonomization has arisen that imposing body of literature going under the name "philosophy of mind," dense with a river of proposals, debates, corrections, deposited into a bibliography that is, by now, boundless, about which Searle, one of philosophy of mind's most representative authors, expressed a clear and pitiless judgment: "the philosophy of mind is unique among contemporary philosophical subjects, in that all of the most famous and influential theories are false" (Searle 2004: 1–2).

Whether or not we share Searle's judgment, it is certainly true that none of the innumerable attempts at a solution has generated consensus or established itself as reliable and preferable. Indeed, often each proposal has been shattered by myriad adjustments and corrections, which are even less effective at giving an adequate response to the problem posed by the discipline. Certainly philosophers of the mind have no lacking in logical or argumentative abilities nor even in metaphorical creativity: the weaknesses undermining the discipline must be sought elsewhere. Philosophy cannot evade this research if it wants to perform the task of understanding and reflection that is proper to it, which includes the commitment to correcting solutions that are inadequate, as well as critically revising assumptions and prejudices, if any.

In the first part of this chapter, I will lay out the principal solutions that have been given for the mind-body problem, to highlight the most relevant theoretical lines in philosophy of mind. This historical part will be followed by an explicitly theoretical-critical section in which I will identify those assumptions that, in my opinion, condition and heavily restrict the research, and which need to be reviewed or, where appropriate, be abandoned. In the third part, I will present the systemic approach, exclusively for its aspects that are useful or significant in philosophy of mind, leaving aside that which, even though of great interest, does not constitute a profitably usable contribution in this specific context. In the fourth and last part, I will advance a proposal inspired by systemic thinking, where the problem of the mind–body relationship will be repositioned in reference to the human being, in his unity, and with the properties that characterize him and differentiate him from other living beings.

The four parts have a certain relative autonomy and can even be considered independently from one another.

The criticisms to standard philosophy of the mind, along with the proposal that has been made here, have greatly benefited from the opportunities for comparison and dialogue with the authors of the essays in this volume, and with the speakers at the numerous systemic seminars that have, for years now, been held at the Catholic University of the Sacred Heart in Milan. I am grateful to all of them for having agreed to participate and contribute to a rare and special experience of interdisciplinary and collective research.

1 First Part: The Status of Philosophy of Mind

Philosophy of mind has taken on the specific task of describing the relationships between mind and brain, or, in more recent times, between mind and body. Philosophers of mind do not work in laboratories, but, like all philosophers, at the desk, where also the results produced in laboratories are utilized for proposing hypotheses sustained by argumentation, ideas, and even metaphors, by means of which to arrive to a meta-level, and therefore philosophical, understanding of the problem under consideration.

In their work of reflection and explanation, philosophers of mind obviously do not start from nothing, but tap into an impressive cultural background, which in part is acquired through knowledge of the philosophical literature and by means of debate within the discipline, and which is, in part, common domain, shared in the historical context in which they operate; this latter is often indicated as "common sense," the vast and incoherent depository of wisdom and superficiality, of profound intuitions, of fragments of scientific and even philosophical theories, often superseded and incompatible among themselves. The task and goal of the philosopher's work is to express general hypotheses and explanations, which, while rooted in the cultural domain embracing them, restore the most complete and coherent vision possible, harmonizing the available knowledge.

Philosophers of mind have worked to carry out this task, taking on the objective of explaining the relationships of the mind and brain. They started from a widely shared concept of mind and brain that would seem to respond to those ordinary experiential data that are irreducible and non-problematic, which can be assumed with some certainty, which are given as follows:

Mind is understood as something that is exclusively accessible in first person: it is private and subjective, equipped with qualitative states called qualia, while the brain and/or body is the biological organ observable in third person, public and objective, characterized by states, dynamics, and processes. This way of considering the mind and brain/body seems, at first sight, sufficiently neutral to represent a good starting point (many would propose to consider this as an "intuitive" or "obvious" piece of data), but if we look closely, it neglects and obscures some facts that are also easily accessible and immediately available, which suggest a general rethinking. The mind is not only private and subjective, but it leaves its mark on the world and transforms it according to its ideas, its choices, its projects; the sciences themselves, both basic and human, are produced by mental activity. The brain/body is not only an object observable in third person, but is also lived and felt in the first person: I see the hand, but I also feel it as mine. Furthermore, and especially, identifying mind and brain as two distinct objects means to undervalue the belonging to, and the dependence of both from, the human being, advancing a dualistic suggestion in the initial project of the discipline.

Although attempts to find an adequate solution to the problem of the mind–brain relationship are countless, it is possible to group them into some—few—influential and shared orientations: dualism, materialism, functionalism.

The dualists sustain that mind and brain are irreducible and in some manner autonomous; at the historical origin of dualism stands Descartes (1641–1642), who decidedly distinguishes the *res cogitans*, caught in self-observation, from the *res extensa*, the object of measurement and science, describing them as distinct and separate substances, having no property in common. Descartes, a tormented dualist, bequeathed to dualists of all ages the same difficulty that he himself could never overcome: if mind and brain are autonomous and heterogeneous, how can we explain the relationships that they, nonetheless, undeniably maintain? Besides this, Descartes left open another problem, even more serious, which in my opinion has heavily conditioned philosophy of mind in an unsavory and subtle way that no one seems to notice, and this is why it needs to be highlighted. The distinction between *res extensa* and *res cogitans*, at least in the Cartesian project, includes in the domain of matter whatever can be expressed in quantitative terms, and attributes to thought the special knowledge only reachable through introspection. The two domains are symmetrical, or so they should be. Under closer inspection, however, they include a fatal asymmetry: matter, opaque and measurable, is inert; it follows laws that it does not know; it knows nothing of itself. Thought, in addition to grasping itself, also grasps matter; it understands it; indeed, it alone may understand because matter, by definition, does not think. If knowledge is one of the properties belonging to thought alone, and thought is excluded from the statute of matter, two unresolvable problems are opened; the first, well-known, asks how thought and matter enter into relation. The second, often not explicitly grasped, but no less serious, opens wide when one wants to find a place for the sciences, that is, for knowledge about matter, within that dualistic ontology; here one comes to the discovery of an unmanageable weakness that undermines dualism, because knowledge about matter cannot be attributed to matter, which does not know, and it cannot even be attributed to the mind, which knows only the world present to introspection. The Cartesian project does not succeed in constructing an adequate ontology; indeed, it proves what it would intend to negate, namely, that ontological dualism is either incomplete or insufficient.

Posterity inherited the problems opened by lame Cartesian dualism and put into act different strategies to overcome them. Some believe that they could escape the problem by denying the mind, but they did not deny enough, in that they held firm the Cartesian significance of "matter" with the consequence of not being able to find a place for mental states, which were not easy to eliminate, obstinately resisting all attempts to cancel them and obstinately reappearing, whether as semantics, or as qualia, or as knowledge. Others mentalized matter, opening to idealism, with the consequent heavy price of misunderstanding the importance of scientific knowledge and the attestations of experience. Still others remained dualists, but the difficulties encountered by Descartes remained to be overcome; indeed, for however unwilling the dualists are to recognize it, the most coherent outcome of dualism is a form of idealism, perhaps even solipsistic: given that knowledge is the exclusive competency of the mind, matter, to itself unknowable, in the moment in which it becomes known, is nothing other than an object of the mind, which ends by dominating without opposition both the ontological and the epistemological horizon.

In view of the serious theoretical shortcomings afflicting dualism, it is difficult to understand its resistance up to present, to be attributed more to its diplomatic virtues than to theoretical soundness: dualism is both conciliatory and resigned. It is conciliatory because it admits the existence of not only physical entities (the brain) but also of mental entities (the mind); and it is resigned because it abandons the study of the brain to the sciences, reserving the study of the mental for philosophy.

The concept of Cartesian matter, with its limits and its criticality, becomes the bearing column of materialism. Materialists sustain that all there is in the world, and which is known or knowable, is matter; what is meant by matter is not entirely clear, but perhaps by settling for a certain vagueness one can limit oneself to considering matter as that which is public, objective, and observable in the third person. If we start from the materialistic assumption, the problem of the mind–brain relation finds a solution simply by negating that the mind is autonomous or independent with respect to the brain. For materialism, only the brain exists, with its neuronal activity, and the difference between mind and brain is only apparent: the mind is a simple epiphenomenon, a phenomenon that is exclusively linguistic, or a folkloristic hypothesis. Materialism is associated with a constellation of complementary theses that support it—scientism, mechanicalism, reductionism, causal closure, causal completeness, physicalism—for which I shall briefly recall the significance.

Scientism asserts that for any problem, the only adequate solution is scientific, attributing the honorary title of "science" to the disciplines working in laboratories; it only recognizes to the natural sciences a capacity of knowledge, thus devaluing the human sciences. It can be observed, quite ironically, that scientism does not stand among scientific affirmations, since it sustains the philosophical thesis that science alone knows; but in so saying, with a rather paradoxical result, it must deny cognitive ability to philosophy. The coherent scientist finds himself in the difficult situation of attempting to uphold a thesis that, being philosophical, does not have a truth value.

Mechanicalism argues that all phenomena and objects of the world are structured by mechanical laws which are the laws of interconnection among objects. From mechanicalism it follows that philosophy, if it wants to know something, must be absorbed by mechanics.

Reductionism is of various types and levels, all of which tend to suggest that objects of the world should be described at the level of microphysics. In philosophy of mind, the reductionist argues that the mind is "nothing other than" the brain.

Those who sustain the thesis of causal closure affirm that only physical causes act upon the world, and therefore the mental has no causal efficacy.

For the thesis of causal completeness, the world is made up of a network of causal connections described by physical laws governing the development of all that happens; every physical event has a physical cause acting according to physical laws, and tracing through these, we obtain a complete explanation of the event.

Physicalism is the joint assumption of the theses of causal closure and causal completeness. It reduces all disciplines to physics, since for the physicalist each event is, and can be, exclusively, a physical event having a physical cause that acts in accordance to physical laws.

Is worth pointing out that all the above-mentioned arguments are not scientific theses subject to some kind of experimental testing, but are philosophical theses which are sometimes made to pass as a "scientific view of the world," whereas it would be more correct to say that some scientific theories are extended in order to structure a philosophical view of the world.

Two comments on materialism.

First, the assertion that reality is made up of matter alone is not susceptible to experimental testing, therefore it is not a scientific affirmation, but rather a metaphysical one, since it sustains the existence of something that is meta-empirical from which the empirical and the experimental—in some manner, however, unclear—derive or have ultimately been formed. The materialist is engaging in a very strong metaphysical thesis, but very few philosophers of mind who claim to be materialists recognize themselves as metaphysicians; indeed, for the most part they believe that materialism is the philosophical position closest to the sciences because it is easily derived from the sciences.

Second, if the mind is nothing more than neuronal activity, there is no one theory that is better or more true than another because theories relating to the mental are nothing but neuronal activity, and neuronal activities are nothing but physical states, which, at most, succeed one another or change: there is no reason to consider one to be preferable to others. With the paradoxical consequence that even if materialism were true, we could not know it, because, if the mental were to fall, everything depending upon it would also fall, including the concept of truth: when we sustain materialism, we are only expressing a neuronal state, just as happens with spiritualism, dualism, animism, mysterianism, etc.

In conclusion, materialism inherits its contradictions from its Cartesian birth, taking from Descartes only one of the dualistic poles, the *res extensa*, undermined by a total opacity to knowledge.

Functionalism seemed to offer a more promising solution to the problem of the mind. Functionalists argue that mental states are functional states determined by causal relationships that occur in the brain. From the metaphysical point of view, functionalism is a mechanicalistic variant of materialism, but with respect to materialism, functionalism describes the mind–brain relationship more accurately and precisely. In philosophy of mind, functionalism has met with the great season of strong Artificial Intelligence, which in the last century has carried forward the enterprise of designing "thinking computers," beginning from the hypothesis that human intelligence is nothing but the ability to manipulate symbols according to given rules, and expecting, as a natural consequence, that a machine capable of carrying out the same operations would be a thinking machine. Enormous funding has flowed into the project of Artificial Intelligence, attracting the most brilliant researchers of various extraction: logicians, computer scientists, mathematicians, engineers, philosophers, dazzled by the brilliant and persuasive metaphor that seemed finally to give an adequate and scientific response to the mind–brain problem: the relationship of the brain to the mind is the same as that of a computer's hardware to its software. The equally imposing disanalogies have not been noticed, or they have been considered irrelevant, and the researchers have not asked themselves if and how the concept of the machine resists when it is used to describe the brain: if we think that the brain is a machine, how should we describe machines so as to include the brain among them? We should also ask ourselves who wrote the human software, namely, the mind. If we take the brain as a machine that functions using instructions, practically isolated from the environment, we are neglecting the fact that not only is the brain not isolated from the environment, but also that it entertains continual relations with the environment, which modify its physical and regulative set-up. The world has been considered equivalent to a tape sending unambiguous signals to the brain, neglecting the strong ambiguity that characterizes objects and phenomena. Functionalism has been faced with objections that by now are well-known, including the mental experiment regarding the Chinese room, using which Searle (1980) proved the irreducibility of semantics to syntax, while, in more recent times, Nagel (2012: 16–20) has shown how no attempt at functional reduction or elimination of the mental can be said to be successful, as there is always a residue of mind, or of knowledge, or of thought, that has no place in a world reduced to functions.

The inadequacies of dualism, materialism, and functionalism clearly show that it is necessary to correct the establishment of the problem, to review its premises, to control its formulation, and eventually drop the conceptual correlatives that support it.

2 Second Part: Tensions and Inconsistencies

Even though the canonical layout of philosophy of mind—with its corollaries and further hypotheses indispensable for supporting it—is almost never directly called into question, concepts and proposals are encountered more and more frequently in literature, introduced almost furtively and as though of necessity, which are incompatible with the previously adopted assumptions. From this tension derive difficulties, contradictions, and unresolved theoretical obscurities.

Let me give some examples.

In order to explain how the mind can be distinct from the brain. Searle (2004: 149) admits the existence of second-level properties, which belong to a system considered as a unitary object. If the mental is considered as a second-level property of a human being, it can be included among the natural properties and incorporated within a strictly naturalistic horizon having no need to open to metaphysical, spiritualistic, or idealistic categories in order to find placement in the mind. Admitting that there are second-level properties is indispensable for sustaining the thesis of biological naturalism dear to Searle, but Searle dedicates few and fleeting remarks to such properties. Let's see if we can understand why. If second-level properties are endowed with causal efficacy, they can intervene in the natural world and produce controllable effects; if they are not endowed with causal efficacy, they are only labels or names, and are insufficient for sustaining biological naturalism, which, when deprived of the mind, must necessarily turn out to be a form of materialism. But Searle has always rejected materialism, so he should admit that second-level properties have causal efficacy; Searle, however, cannot make this move, because he remains faithful to microphysical physicalism, according to which only those physical forces have causal efficacy, which govern the behavior of the particles making up the natural world. The tension between physicalism and autonomy of the mental remains unresolved and clearly shows that we must renounce one of the two in order to resolve the problem of the mind–brain relationship. Searle's attempt, though remaining a half measure, should still be appreciated because by contrast with the mainstream, he clearly recognizes the impossibility of finding a place for the mind in materialism; his attempt to find a place for the mind in nature cannot be called a success, but it has the merit of not eliminating mental states from world phenomena: making recourse to reductions is not a good philosophical explanation of a problem, is only an over simplification.

Furthermore, Searle (2004: 118–123), always in order to identify a conceptual place where some sort of independence of mind from body can be admitted, introduces the distinction between causal reduction and ontological reduction. One phenomenon is causally reducible to another if it can be completely explained in terms of the other, but as Searle observes, causal reduction, of itself, does not include ontological reduction, which consists in not having causal properties other than those acting at the microlevel: explaining what the cause of a phenomenon is does not imply that the phenomenon is nothing other than its cause. By applying such a distinction to consciousness, Searle concludes that even if a causal reduction of consciousness were possible, by explaining it as neuronal activity, there would still remain a need to explain consciousness as the phenomenon only accessible in first person. Searle makes it quite obvious that an object is other than the procedure for obtaining it, and that in order to explain it, it is necessary to place oneself at a different level of understanding, but from this correct affirmation, he does not draw the inevitable conclusion that the explanation of the world in microphysical terms is incomplete since the world is populated by objects that are ontologically irreducible to microphenomena. If consciousness is ontologically irreducible to the brain, as Searle asserts, the physicalist thesis is false and must be abandoned. But Searle does not abandon physicalism, inexorably compromising the attempt to recognize an autonomous ontological status to the mind.

Crane (2001a, b) takes the step, not completed by Searle, of attributing causal capacity to the mind, accepting the consequence of renouncing physicalism. Crane sustains that there are "emergent" properties that are irreducible to the physical and endowed with causal powers, from which it follows that the physical is incomplete and insufficient for describing the properties of all the objects in the world. Emergentism is exploited by Crane in order to resolve the mind–brain problem: the mind emerges from the brain, and this is the final level of understanding available to us, which is to be accepted as a fact of nature, without being able to figure out which type of relationship exists between brain and mind. Crane succeeds in overcoming physicalism—the strongest conceptual obstacle to attributing at least a possibility of autonomy of the mind—but leaves open to further and deeper investigation the concept of emergence and its consequences, among which we can recall: the concept of downward causation, to which Crane (2001a: 63) dedicates a sober note; the inevitable abandonment of epistemological reductionism; the admission of an ontological pluralism. Emergentism is shared by many scientists, including Libet (2004: 162–163), Edelman (2004), and

Damasio (2010), who attempt to make it coexist with materialism, the last bastion of the so-called scientific view of the world.

Others, including Chalmers (1996) and McGinn (1989), may properly be defined as mysterians because for them the problem of the mind, even while not reducible to neurobiology, remains, of itself, mysterious. In this case, the impossibility of explaining the mind depends on adherence to that strong form of scientism averse to admit humanities in the domain of "sciences." Scientism, in arguing that only the natural sciences may give a valid solution to any problem, is forced to consider as "mysterious" those problems that cannot be treated by such disciplines. This position has the undoubted merit of not denying, reducing, or eliminating consciousness, but is not open to solutions coming from other disciplines since it does not go beyond scientism.

Nagel (2012: 35–47) expresses with great lucidity and effectiveness the incompatibility between autonomy of mind and completeness of the sciences. By contrast with the mysterians, for Nagel the mind should not be considered a mystery, but rather a problem that the sciences have failed to face and that, in order to be addressed, requires a great conceptual demobilization. We must abandon reductionism, naturalistic materialism, mechanicalism, and even scientism, which have shown themselves incapable of explaining facts present in nature such as the mind, intentionality, knowledge, and we must introduce alternative conceptual instruments that make possible a systematic understanding of the natural world. Among these, Nagel considers the admission of teleological principles to be essential, and indispensable in accounting for the historical development of the world as an ordered whole. Nagel looks confidently at the possibility of activating other models of understanding that will be able to overcome those limits of the dominating models, which he has so clearly highlighted.

Even without engaging directly in philosophy of mind, Dupré (2001) makes a remarkable contribution to that conceptual reorganization desired and initiated by Nagel, not only with his criticism of causal completeness—and therefore of determinism—but also with his openness to pluralism in epistemology, in the sciences, and in ontology. Determinism, says Dupré (2001: 163), is a metaphysical hypothesis that is a "philosophical free rider on the scientific world view," devoid of empirical support. The causal order, rather than being general, as determinism would purport, is partial and incomplete all around. The fall of determinism makes it necessary to revise the concept of linear and efficient cause supporting it, and Dupré does not pull back, introducing an important critical reflection on the concept of cause dominant in the philosophical horizon from Hume (1739–1740) to Mackie (1974). To overcome the limited efficacy of such a concept of cause, Dupré usefully suggests to introduce the concept of downwards causality. The empirical world, no longer constrained in the rigid determinism of Laplace's great machine, also escapes microphysical reductionism: causal efficacy is also attributed to structural levels superior to that of the microphysical, with the consequence that the objects that are formed through processes of integration of lower-level unity acquire different, but equally real, causal properties than the lower-level objects. A world populated by objects with different and irreducible properties entails an epistemological pluralism, which introduces into the panorama of knowledge a plurality of forms of knowledge, the "sciences," each with its own object and its own criteria of validation and description.

Current proposals in philosophy of mind have been further weakened by advancing research in the neurosciences, which have enriched our knowledge of the brain radically modifying the biological references adopted by philosophers of mind. The identity of cerebral and mental states, that supports the type or token identification between mind and brain, has been refuted: numerous researches have proved a clear asymmetry between mind and brain, as documented by the observation that only a part of neuronal activity is available to the mind (Frith 2007); to this is added the data, now well-established, that the brain cannot function except in connection with the entire body (Damasio 1994). Other studies have proved that there are activities in which the mind operates autonomously with respect to bodily states, correcting neuronal distortions, organizing vision in accordance with cultural expectations, habits, models (Gregory 2009), even placing a veto on biological tendencies.[1] Numerous and intertwined connections between mind and brain are also documented: the placebo effect (Colloca and Benedetti 2005) indicates an influence of the mental over the body, while bodily states alter what mental consciousness experiences. The functionalist thesis, according to which the brain is a machine, is weakened by the description of the brain as a plastic organ, transformed by experiences and individual history (Edelman 2006), while autonomy of the mental, the way dualism puts it, is falsified by Damasio's research, who has proved that the mind is never "pure," but is constantly impregnated with corporeality and emotions. Many neuroscientists are oriented by their research to consider body, mind, and brain as a unitary object of interacting parts, bringing attention to the human being in its unity and integrity. This tendency is difficult to reconcile with the materialism often admitted to by neuroscientists themselves, but as we know, materialism is a metaphysical thesis, which must be checked philosophically; when a neuroscientist claims to be materialist, she is not making a scientific affirmation, susceptible to at least one experimental project that might falsify it, she is simply declaring what sort of metaphysics she personally adheres to outside the laboratory, in her free-time thinking, for which her authority as a scientist cannot be asserted as proof.

The panorama in which philosophy of mind moves, to seek and propose its responses, is populated by conceptual ghosts that prevent a real progression of the discipline (the various aforementioned—isms), and which have infiltrated both scientific research programs and what scientists assert publicly, generating inconsistencies, misunderstandings, difficulties. I hope I have demonstrated sufficiently that philosophy of mind does not change on its own, because, more than other philosophical specialties, it depends strictly on data taken from observations and experiments, which can only derive from scientific disciplines; and because its object, as present in the empirical world as it is in the mental, does not permit formal reductions or solipsistic mentalizations that would sacrifice the empirical, but does not allow itself, either, to simply be absorbed into scientific results where the mental has no place.

[1] Libet (2004: 137–150), Libet hypothesizes that a *conscious veto* can control whether an action initiated unconsciously in the brain will or not take place.

In order to carry out that modernization of the world view needed to make philosophy contemporary of its time, the first contribution philosophy can, and must, give—and I hope the preceding pages have given at least some inspiration for this—is, certainly, a rethinking of the conceptual instruments that, while admitted as valid in the past, today appear clearly insufficient for understanding the richness and novelty of the data available; but a second and equally important philosophical commitment is that of introducing approaches and methods that may better satisfy the renewed needs for explanation.

3 Third Part: Contributions of Systemic (and Non-systemic) Thought Useful in Philosophy of Mind

What available tools do we have for making more clear, precise, and detailed a perspective recognizing the world as a rich plurality of objects that are distinct as individuals and different by kind, closely linked by dynamics of alternation and substitution, and, for some of those entities, of birth and death?

Once we have abandoned the analytical attitude, for the well-known reason that it ends up transforming individual objects into unmanageable ghosts, a promising perspective is offered by the systemic approach, capable of describing the relationship between an object (on whatever scale and to whatever context the object belongs) and its parts without having to pay the price of reducing the object to nothing but its parts.

According to systemic thinking, an object should be described as an organization of parts connected by relations, and having properties that the parts do not have; such properties are called "emergent," or "second-level," or "systemic" properties. A common feature of all objects considered as systems is the primacy of organization over parts. The organization consists of a network of relationships that bind and connect the parts to each other, obtaining the result of stabilizing the object and guaranteeing its unity and identity; the organization, which is constituted exclusively of relationships, is not directly observable and can only be traced back, by an "intelligent" observer, starting from an understanding of the effects that it produces, namely, the object's behavior. While a high level of stability characterizes the organization, the parts can be replaced—without compromising the identity of the system—with other parts that satisfy the requirement of compatibility with that organization. It should be noted that the parts do not remain inert with respect to the organization they enter: they react to the imposing of systemic constraints with a certain adaptation: this phenomenon has been observed and described already in physics and in biology.[2] Emergent properties, which support the identification and distinction of objects, are to be discovered (that's why they are said to be "new" or "unexpected") as a result of an observational focus, which detects them and makes them accessible.

[2] Alessandro Giuliani proves in many works that cells undergo change depending on the biological context in which they interact. See Kohestani et al. (2018).

Although the term "emergence" can induce the idea that such properties are entirely derivable from system constituents, the concept of "emergence" sustains precisely the opposite argument, namely, that the emergent properties cannot be derived from the properties of the elementary constituents. The concept of emergence highlights the fundamental nature of the phenomenological moment in which objects and their properties are observed and, in so doing, it recognizes the relevance of the observer, with the cognitive capacities proper to him, in intercepting and understanding those phenomena (Crane 2001a, b; Minati et al. 2008).

It is clear already from these first remarks that systemic thinking does not attempt a formal reconstruction of the world, but starts from a plan of observation in which, as a function of the cognitive availability of the observer, different objects endowed with properties rendering them identifiable are found. After this moment of selection and observation, there must follow a moment of theoretical comprehension, provided by an adequately rich and faceted instrumentation. Systemic thinking implements numerous concepts functional to understanding of the *explicandum*, identified by the question: what constitution is to be attributed to a certain object to explain its behavior? *Explicantes* of the constitution of the object are the organization, the parts, and the systemic properties, which have been mentioned above. To explain the behavior of an object, as required by the *explicandum*, is necessary to introduce, or to specify, additional concepts that can be seen as capable of considering the temporal factor in which the behavior is realized, which is nothing other than a change over time. To this end, concepts of structure and environment are used, further pillars on which systemic thinking builds its idea of the world. By structure, we mean the values of the variables an object presents at time *t*, thereby assigning to an object a range of variation of the parameters within which identity is safeguarded.[3] With the concept of structure, we are capable distinguishing between the way a system presents itself to the individual act of observation carried out at a certain time, and other values of variables which that object can also assume, allowing us to identify an object as "the same" though not identical, within the dynamic process of states allowed by the repertoire of variables that characterize it. The environment is the basin, itself mobile and flexible, in which the dynamics of the object are realized, often activated by external, meaning environmental, pressures, to which the system reacts by means of continual acquisition and releasing of constraints and parts. Despite the vast array of relationships that bind a system to the environment, it is possible to find a boundary, however mobile and approximate, between system and environment, depending on the density of constraints, which within the system is greater than that which binds the system to its environment (Hooker 2011: 868).

The concepts of organization, parts, emergent property, structure, environment: these are the fundamental references of the systemic approach, sufficient however to delineate a different vision of the world: the systemic world is not a fixed, stable,

[3] The term "structure" is used in literature with two rather different meanings. In ontology, transferred from logic, structure is understood as the correlation between parts that identifies a set, while Maturana and Varela (1985) understand it as the values of the variables taken on by an organization at time *t*, in dynamic perspective.

and closed environment, to which time is added as one more complication to take into account, but it is an intrinsically dynamic world, in which, thanks to relations of interaction and interference of systems among themselves and with the environment, continual phenomena of emergence of objects and properties take place. Stability and change enter into this vision not as two separate or concurrent conceptual poles that contest the "final" physical or metaphysical explanation of the world, but jointly indispensable in rendering ontologically possible and humanly comprehensible that which we observe. With a metaphysical undergirding of the anthropic principle, the question about the world becomes: Given that the world is as it is, composed of stable objects and dynamic ensembles, what hypothesis needs to be introduced in order to render it comprehensible? The systemic approach, by suggesting that our world is realized owing to a simultaneous array of stability and transformations, of ordered structures and ruptures of symmetry, of acquisition and abandonment of constraints, opens the way to the theoretical commitment that is continually and closely related to experience.

For further details on the constellation of concepts involved in the systemic approach, the reader is referred to the systemic literature, which, already very vast,[4] is continuously being enriched with contributions, debates, and proposals beyond the possibility of even being mentioned here.

In the following, I will limit myself to expanding and discussing only those concepts developed in systemic thinking, which appear useful in philosophy of mind, because they restructure its references or correct some of its current distortions.

With the systemic approach, a hierarchical perspective is introduced, both internal and external to objects: a system is constituted through the correlation among parts, placing it at a certain ontological level: that in which the object, with its properties, is detectable. The parts constituting the object are positioned at a level inferior to that of the object, a level we can call subsystemic. The system initially considered can, in turn, enter into larger organizations, which we shall call suprasystemic, in which it takes on the role of a part. If we consider a human being and read him in the perspective of systemic hierarchy, we will be able to attribute to him properties of the systemic level, for example, language, but also parts of the sublevel, such as internal organs, which in their turn, may be described as systems; furthermore, the same human being may enter as a part into supra-systems that constrain his behavior, such as family, work environment, society.

A proper hierarchy highlights the fact that at each level there appear properties typical of that level and invites to refer them to the objectual level to which they actually belong. The correct attribution of a property to the systemic level supporting it becomes decisive for understanding many phenomena. Errors of level, very frequent even in the sciences, generate distortions and misunderstandings, with serious repercussions that may even be practical. Bertolaso (2018) has exemplarily highlighted an error of level committed in cancer research, which has concentrated its impressive efforts on cells, neglecting tissues, the actual level on which cancer manifests, and where it ought to be sought.

---

[4] For an excellent bibliography, see Minati et al. (2016).

From the hierarchical distribution of objects comes a consequence that is worth stressing: there are no closed systems, either from the ontological point of view or from the epistemological point of view. Given that every system participates in different ontological levels, both higher and lower, there is no one level at which the object is ontologically contained and enclosed, with the consequence of excluding that one may look for the "best" or "true" level at which to consider it. It is also eliminated the possibility, even if only hypothetical, of complete knowledge of an object because the relations it entertains with other hierarchical levels, both higher and lower, involve that the knowledge of each level, constitutionally open to the others, can only be incomplete. What follows is an unpredictability that we may call objective, in that it does not depend on our cognitive limits (in this case, it would be a subjective unpredictability), but on the way the world is structured, as an array of open hierarchies that (objectively) possess and manifest (to the cognitive entity capable of grasping them) new properties, whether ontologically or epistemologically. The environment, not only physical, becomes the site of systemic dynamics that are structurally open, also, thanks to hierarchical interactions.

From the two concepts of hierarchy and environment descend important consequences on the concept of cause. If, for "cause," we understand, with Aristotle, any valid explanation of a phenomenon, or, in more current terms, anything that produces a variation in the behavior of an object, we must consider as bearers of causal efficacy not only the so-called efficient cause, but also the rules that constrain behaviors from above (*top-down* or *downwards* causality), as also the pressure exerted by changes that take place in basic constituents (*bottom-up* causality).[5] Top-down causality must be introduced when it becomes necessary to unify a plurality of behaviors that would remain incomprehensible if one tried to understand them by resorting only to the two concepts of chance or probability. Let us imagine we want to explain the behavior of four card players; if we use only the concept of efficient cause, integrating it with concepts of chance and possibly of probability, we will say that the players are the (efficient) cause of the behavior of the game, explaining the frequency with which a card (and hence all cards) appears on the table as a function of probabilistic calculations. If these calculations are correct, we will make affirmations that are certainly true, without having thereby understood more than a very little of what is happening on the playing table. Our understanding becomes more refined if we introduce the concepts of bottom-up and top-down causality; using the first, we will explain how a pack of cards should physically be composed so that the players can make their moves; using the second, we will find the rules that govern the game, from which derive the players' behaviors in their selection of cards to be played. Clearly, the two forms of causality act together; if there is not a pack of cards composed in the appropriate manner, the rules have no field of exercise, and if there are no rules, there is no game; therefore to understand the phenomenon under consideration, that is, a game of cards, we must keep in mind their simultaneous causal interlinkage, from which derives the game that effectively is realized in

[5] *Top-down* (or *downwards*) causation is presented and discussed in: Anderson (1972), Laughlin (1999), Auletta et al. (2008), and Ellis et al. (2012).

a certain period of time, and which becomes "this game," in which the rules are respected and an appropriate pack of cards is available. The interlinkage of the two forms of causality may be indicated as middle-level, or middle-out, causality, which derives from the configuration of the relationships among the elements of a system. With Aristotle, we could call it "formal causality;" systemically, we can indicate it as the causal efficacy of the organization. Biology is nowadays introducing a further concept of cause, called "causality by absence," which can advantageously complement the precedents, and which awaits a proper conceptualization.

By coupling efficient causality with other concepts of cause, systemic thinking makes it possible to understand certain characteristics and behaviors of the objects that remain otherwise unexplainable, with the consequence that the concept of cause becomes an analogous or pluralistic concept, acquiring that flexibility which makes it suitable for describing a world of intricate and multiscale influences, with constant changes of systemic variables and properties, losing the rigidity and explanatory poverty of an unambiguous reduction.

Argumentation, as well, which traditionally has guaranteed knowledge through the canonical deductive and inductive procedures, can include the useful addition of abduction as an epistemological instrument endowed with high cognitive efficacy in the comprehension of phenomena and objects of the world. Abduction, as described by Peirce (1931–1958), consists in finding or discovering hypotheses—that can be both particular or universal—which are useful in explaining a fact or set of facts within a context of incomplete information (Frixione 2007: 6–9, 123–129; Urbani Ulivi 2016). While it is always possible to build formalized models by providing them with a complete context of information, in most cognitive problems we only have incomplete information, and of necessity general or particular hypotheses must be introduced to over-determine the available data. If, by the admission of abduction among the effective argumentative proceedings, we are compelled to renounce the ideal of closed, certain, and complete knowledge, the episteme of Platonic memory, this loss is compensated for by the increase of understanding of the activity of human cognition, to which a capacity is attributed and recognized that can be qualified as "creative," which achieves cognitive objectives by utilizing procedures that cannot be formalized by means of given rules. Once recognized that knowledge is open, uncertain (Licata 2012), and incomplete, the claim that there exists one and only one "true" knowledge also falls, even if only as an ideal to be pursued. In a systemic view, abduction is used to discover and identify second-level properties that cannot be derived deductively from the parts, that need to be introduced to explain behaviors or objects. A classic example is the attribution of free will to human beings, which is not directly observable, nor deducible with logical necessity from the human constitution, but which is introduced with an abductive argumentation to explain human behaviors which would remain opaque to any attempt at deduction. Abduction, by contrast with deduction, which allows for the construction of exclusively formal systems, roots the logic in the world by reevaluating the importance of an accurate and refined phenomenological description, which is also realized with attention to scientific results that enter not only as facts, but also by right, into both the real and cognitive human world.

I would like to emphasize the importance in systemic thinking of the phenomenological moment, which observes the objects of the world and identifies them by distinguishing their properties: the link with reality and its attestations is irreplaceable for the systemic observer, who knows he needs to discover objects and properties in the world, and who in this work utilizes all the available scouting tools, from common sense, to science, to philosophy, in an effort, first of description, then of understanding, in which it is up to the world to pronounce the last word.

The last contribution of systemic thinking usable in philosophy of mind is the immediate openness to interdisciplinarity, which naturally follows from the recognized openness of each system to interactions with multiple levels and with the environment. If we consider the scientific disciplines as cognitive systems, to each one is accorded that autonomy derived from having an object of its own, but also the openness to cognitive references that originate from the cultural environment in which it is immersed. The interdisciplinary relationships contribute to the progress of each individual discipline (or science, if you will) enriching the conceptualization of the available object also within a certain discipline.

The scenario of the systemic concept I have outlined thus far is far from being complete, but it is the most promising track to follow and work on, in order for philosophy of mind to be able to rewrite its statute, without the heavy claims it has thus far taken on from the cultural universe within which it has operated.

4 Fourth Part: Proposal for a "Systemic" Philosophy of Mind

In the contemporary world, a process has been launched on several sides deconstructing physicalist triumphalism, now undermined by the progresses of biology, which is discovering the irreducibility of its object to physics and which demonstrates the inadequacy of the analytical method alone for grasping the laws governing the living. If as a natural consequence reductionism falls, then also the physicalist theses of causal completeness lose its epistemological supports, and the mechanicalism can no longer be credited as a general vision of the physical world, but at most, as a local description, suitable for representing certain phenomena. The space opens to support a pluralistic ontology which, in turn, leads to a pluralistic vision both in epistemology and in the sciences: if the world is populated by different objects that cannot be reduced to one single level, then different approaches and sciences using different methods and criteria for describing those different objects are not only to be tolerated but become indispensable. Dilthey's (1914) fracture between natural sciences and sciences of the spirit no longer has reason to exist, in that the human sciences also naturally find their place among the plurality of sciences, as they describe objects that are inaccessible to the so-called natural disciplines, though this does not make them any less real.

This impressive transformation also invests philosophy of mind, not only by imposing revisions, but also by opening unexpected possibilities.

The first question to ask is whether the problem instituting the discipline—namely, Which is the relationship between mind and brain?—is to be accepted as it is, or whether it needs to be reviewed. Its traditional formulation implies, in a certain manner, that mind and brain are two separate and distinct objects, for which it is necessary to investigate the relationships that doubtlessly exist (and why?), but which at first sight are obscure. And this does not surprise if, as has been said, it was Descartes who established the problem, giving birth to a dualistic assumption that has been largely accepted and adopted in the successive philosophy of mind. If the dualistic formulation is dropped, attention can finally be moved to the human being as a unitary entity, which has been, up to now, notably absent from the post-Cartesian horizon. Mind and brain do not flutter freely throughout the world, and they must be brought back within that unified ontological horizon, the human being, in which they are found and in reference to which, only, there may be sense in posing the question of their relationships: the mind is the human mind, the brain is brain—and body—of human beings, and their relationships take place through the inherence of both within the unitary human subject sustaining them.

The theme at this point becomes: What relationship is there between mind and brain in human beings? The correction is of no small account because it shifts the search bar from two objects, mind and brain, to one unitary object, the human being, by identifying a new and different task, namely, to explain what relationships the mind and brain maintain with the human being, before explaining that maintained with one another.

First, it is necessary to attribute an ontological place to the human being: it is an object among the others of the world, as experience undeniably affirms, and it has properties that other objects do not have, which differentiate it and make it identifiable, as experience affirms, as well. It has been seen above that if we consider an object as a system, the complex interweaving between the parts becomes clarified, without thereby requiring reduction of the objectual unity to its constituents. Thinking in systemic terms about that special object given in our experience which is the human being, we would describe it as an integral and unitary phenomenon that is structured and specified owing to the continuing and complex interactions with the multiple properties and subsystems characterizing it, and with the supra-systems and environment with which it relates, with process dynamics of reaction and adaptation.

Now let us consider mind, brain, and body, the problematic terms of philosophy of mind, bringing them into the systemic context.

With respect to the human system, the mind is a second-level property (we can also call it emergent or systemic), which pertains to the human phenomenon in its unity. As a systemic property, it must not be sought in the parts of the system, because it does not appear in these, just as architecture does not appear if we take into consideration the bricks of the building, because that is not a property of the bricks, but of the entire building. The mind as a systemic property has causal efficacy on the entire system, in that it contributes to instituting, maintaining, and recuperating the systemic relationships, both internal and with the environment, acting as potent factor of *top-down* causation. The pressure exerted by the mind on the subsystems and also, to a lesser degree, on the supra-systems, modifies both the former and the latter, adapting them to integration and relationships within the human system. Once clarified that the mind pertains to the entire human system, obviously nothing inhibits us from shifting attention from the human being to the mind, by studying its characteristics and qualities, as a property or as a phenomenon in itself. Just as the linguist can study style or syntax or phonemes as phenomena that can be isolated for the purposes of description, notwithstanding the fact that in linguistic practice, for example in a literary text, they are present together. If we concentrate on the mind, we will be able to grasp and describe the traits that make it special and that have always attracted the attention of philosophers: it not only makes the private, subjective world accessible in first person, it also makes available the public, objective world to human, and scientific, inquiry. The mind affixes its seal in science, as well, as Edelman (2006: 156) clearly sees: "Science is imagination in the service of the verifiable truth." This conception finds its natural and powerful antecedent in Aristotle, who indicated the *psyché* as principle of human activity, that which characterizes its nature, impregnating the *bios*, onto which it hinges ontologically. The possibility of focusing attention on the mind induced Descartes into the error of attributing to it an ontological autonomy, considering it as a *substance*, endowed with its own substantial independence, free creator of worlds, indifferent to the bond with corporeal reality. In the systemic conception, the relationship between body and mind is held solidly firm, in that both belong to the human phenomenon.

If, as we have seen, the mind is an emergent property (or an emergent phenomenon) how is the body to be understood?

It should be kept in mind that, with regard to the body, as well, there is not one and only one point of view to favor, in seeking a perspective that could return it fully theorized, but there are different points of observation, each with a more or less refined apparatus for proof and documentation, which take into consideration different aspects, capacities, and functions. But—and in this regard, the systemic approach bears an indispensable clarification—it is important to avoid errors of attribution and level. By limiting ourselves to just two variants, we can focus on the body by examining its biological properties, as *bios*, or by pointing out the characteristics owing to which it is *a human body*. The *bios* body is the object of the biological disciplines that study its properties, laws, and behaviors as the body of a living being. Remaining within this context, the *bios* body may be studied as a system, as *systems biology* effectively appears oriented toward doing, but the level to which the investigation attests is, and remains, biological. If the question turns on: How to think of the body in relation to the human being?, the answer must be situated at the level of the human global unity, for which the human body constitutes the observable phenomenon; in this case, the body is the object identified as the system, just as we find it in the world, an organized object which has a behavior in time and which is identifiable, thanks to the persistence of emergent properties, which cannot be reduced to the nature of its elementary components. It constitutes the phenomenon available to observation, the understanding of which proceeds with the concepts refined by systemic thinking and with the abductive inferences that are indispensable for understanding its emergent properties and behavior.

If, then, the question were not related to the body, but to the brain, the answer would be simpler; the brain is a subsystem, or part, or element, if you will, as much of the *bios* body as it is of the human body. As a part, as for all parts, it manifests its integration into the human system by acquiring "humanized" characters, both biological and systemic. Obviously, it can be studied as an object in itself, and become the focus of neurosciences, which must pay close attention to avoiding errors of level, such as, for example, attributing to the brain properties that are not of the brain, but of the entire human unit, and they must also take into account that a complete explanation of their object is impossible, because that object, the brain, is structured and specified also through the interaction it maintains with the body, with the environment, and with other objects that go beyond neurological description.

If the mind is a human systemic property and the body manifests the system over time, what relationships do mind and body maintain? Or, in other words: How can one respond to the traditional problem of philosophy of mind while assuming a systemic perspective?

First, the question must be formulated differently, so as not to introduce an error of level: What relation do mind and body have in reference to that unitary phenomenon which is the human being, in which they can be found, and what relationships can, as a consequence, be attributed to them? The answer is that the (human) mind is a property of second, or systemic, level of the human being, while the (human) body is the human system as it appears to phenomenal observation; and this, since it is detectable in time t, manifests the structure—in the sense of the values of the variables—assumed by the human system in that given time. If the observation is extended in time and the instantaneous value of time t is brought back to within extended temporal dynamics, then the human system with the variables allowing its alterations over time must be understood as a process subject to change, capable of assuming different behaviors over a span of time. An answer to the question "What relationships are there between mind and body?" must first specify whether we are referring to the body as *bios* or to the body, properly and exclusively, of the human, because in order to answer, we must position ourselves at the level and in the perspective in which the question was asked. If the body is understood as *bios*, the question is asked at a subsystemic level with respect to the human, a level in which the mind does not appear: in this case, body and mind do not maintain relations because the mind goes beyond the *bios* level.

If, on the other hand, the body is considered at the human systemic level, then the relationship established between mind and body is that existing between system and emergent property: the system sustains the property, and the property interacts with the system, in a development in which the two levels of causation, *top-down* and *bottom-up*, express the behavior characteristic of that system, in the resulting *middle-out* causation.

The two responses do not compete with one another: there is not one that is better, or more adequate, nor one that is worse, or weaker. There simply are two different planes, and each has advantages and limitations. If we are at the *bios* body level, we know that we cannot and ought not seek the mind; if we are at the human level, there the mind will appear, as a global property of the system. Therefore, to resolve the problem of philosophy of mind, we need to take into consideration the unitary, complex, and organized human phenomenon in which the mind manifests; whereas if the mind is sought on other levels, such as the neurological or purely biological, meaning in subsystemic levels, the mind simply disappears; and this explains the difficulty of philosophy of mind, and also of neurosciences, in finding the mind, when restricting the horizon of research to the *bios*.

The systemic approach has allowed a reformulation of the mind–body problem within a new and different perspective, thanks to which it is possible to identify and delineate a solution, at least in general lines. This is a proposal that needs to be refined and articulated, but which already demonstrates the capacity to provide a foundation for rethinking a constellation of other concepts and problems, including: it is possible to attribute not only conscious activity, but also unconscious activity to the mind; creativity can be recognized as the capacity to grasp and institute new relations, opening to art; disease and health can be understood in relation to the harmonic integration among parts and levels; liberty can be recognized as one of the systemic properties of the human, without thereby conflicting with deterministic descriptions that can easily be recognized for explanatory efficacy in highly limited areas and problems; self-awareness should be seen as one of those systemic properties that allows human beings to think of themselves within a world scenario, and which depends upon having a mind.

If, then, the inquiry, called for by the systemic rewriting of the mind–body problem, would open itself to philosophical anthropology, considering the human being in its entirety and its environment, it is now clear that the understanding of a complex object such as the human being, whose description cannot be reduced to a single formal model, must take into account the contributions that the many and various disciplines can attain and exchange with one another. The systemic approach, intrinsically open, presses for the various contributions, with the awareness that all are necessary, that none itself is sufficient, and above all in the conviction, inscribed in its statute, that the wealth of processes and emergent properties characterizing our world and ourselves cannot be grasped except through continual interaction and integration of "objective" experimental data and "subjective" theoretical understanding, converging to bring about the world as seen by us, our world, in which we are both actors and spectators.

References

Anderson, P. W. (1972). More is different. *Science, 177*, 393–396.

Auletta, G., Ellis, G. F. R., & Jaeger, L. (2008). Top-down causation by information control: From a philosophical problem to a scientific research program. *Journal of The Royal Society Interface, 5*, 1159–1172.

Bertolaso, M. (2018). *Philosophy of cancer. A dynamic and relational view*. New York: Springer.

Chalmers, D. (1996). *The conscious mind*. Oxford–New York: Oxford UP.

Colloca, L., & Benedetti, F. (2005). Placebos and painkillers: Is mind as real as matter? *Nature Reviews Neuroscience, 6*, 545–552.

Crane, T. (2001a). The significance of emergence. In C. Gillett & B. Loewer (Eds.), *Physicalism and its discontents* (pp. 207–224). Cambridge: Cambridge UP.

Crane, T. (2001b). *Elements of mind. An introduction to the philosophy of mind.* Oxford: Oxford UP.

Damasio, A. (1994). *Descartes' error. Emotion, reason and the human brain.* New York: A. Grosset–Putnam Books.

Damasio, A. (2010). *Self comes to mind. Constructing the conscious brain.* New York: Pantheon Books–Random House.

Descartes, R. (1641–1642). *Meditationes de prima philosophia* (2nd Ed.). Paris: M. Soly. Amsterdam: L. Elzevirium.

Dilthey, W. (1914). *Einleitung in die Geisteswissenschaften. Gesammelte Scriften* (Vol. 1). Leipzig–Berlin: Teubner.

Dupré, J. (2001). *Human nature and the limits of science.* Oxford: Oxford UP.

Edelman, G. M. (2004). *Wider than the sky. The phenomenal gift of consciousness.* New Haven and London: Yale UP.

Edelman, G. M. (2006). *Second nature. Brain science and human knowledge.* New Haven and London: Yale UP.

Ellis, G. F. R., Noble, D., & O'Connor, T. (Eds.). (2012). Top-down causation: An integrating theme within and across the sciences? *Interface Focus, 2*, 1–3.

Frith, C. (2007). *Making up the mind. How the brain creates our mental world.* Hoboken, NJ: Wiley.

Frixione, M. (2007). *Come ragioniamo.* Laterza: Roma–Bari.

Gregory, R. L. (2009). *Seeing through illusions.* Oxford: Oxford UP.

Hooker, C. (2011). Introduction to philosophy of complex systems: B. In C. Hooker (Ed.), *Philosophy of complex systems* (Vol. 10, pp. 841–909). Amsterdam: Elsevier.

Hume, D. (1739–1740). *A treatise of human nature.* London: Noon.

Kohestani, H., Totonkuban, M., Di Paola, L., Todde, V., & Giuliani, A. (2018). The basic principles of topology-dynamics relations in networks: An empirical approach. *Physica A: Statistical Mechanics and its Applications, 508*, 584–594.

Laughlin, R. B. (1999). Fractional quantisation. *Reviews of Modern Physics, 71*, 863–874.

Libet, B. (2004). *Mind time. The temporal factor in consciousness.* Cambridge, MA–London: Harvard UP.

Licata, I. (2012). Seeing by models. Vision as adaptive epistemology. In G. Minati, E. Pessa, & M. Abram (Eds.), *Methods, models, simulations and approaches towards a general theory of change* (pp. 385–400). Singapore: World Scientific.

Mackie, J. L. (1974). *The cement of the universe.* Oxford: Oxford UP.

Maturana, H. R., & Varela, F. (1985). *Autopoiesis and cognition. The realization of the living.* Dordrecht: Reidel.

McGinn, C. (1989). Can we solve the mind-body problem? *Mind, 98*, 349–366.

Minati, G., Pessa, E., & Abram, M. (Eds.). (2008). *Processes of emergence of systems and systemic properties. Towards a general theory of emergence.* Singapore: World Scientific.

Minati, G., Pessa, E., & Abram, M. (Eds.). (2016). *Towards a post-Bertalanffy Systemics.* New York: Springer.

Nagel, T. (2012). *Mind and cosmos. Why the materialist neo-Darwinian conception of nature is almost certainly false.* Oxford: Oxford UP.

Peirce, C. S. (1931–1958). Upon Logical Comprehension and Extension. In: *The collected papers of Charles Sanders Peirce* (2643 ff.). Harvard: Harvard UP.

Searle, J. (1980). Minds, brains, and programs. *The Behavioral and Brain Sciences, 3*, 417–457. (Reprinted from *The Mind's I*, pp. 353–372, by D. Hofstadter & D. Dennett, Eds., 1982, New York: Bantam Books).

Searle, J. (2004). *Mind. A brief introduction.* Oxford: Oxford UP.

Urbani Ulivi, L. (2016). L'abduzione come momento della scoperta in contesti di realtà. Abduction is the inference that discovers a solution in problematic contexts. *Cassazione Penale, 56*, 2240–2251.

MIX
Papier aus verantwortungsvollen Quellen
Paper from responsible sources
FSC® C105338

If you have any concerns about our products,
you can contact us on
ProductSafety@springernature.com

In case Publisher is established outside the EU,
the EU authorized representative is:
Springer Nature Customer Service Center GmbH
Europaplatz 3, 69115 Heidelberg, Germany

Printed by Libri Plureos GmbH
in Hamburg, Germany